RHEOLOGY: AN HISTORICAL PERSPECTIVE

RHEOLOGY SERIES

Advisory Editor: K. Walters FRS, Professor of Applied Mathematics,
University of Wales, Aberystwyth, U.K.

Vol. 1 **Numerical Simulation of Non-Newtonian Flow**
(M.J. Crochet, A.R. Davies and K. Walters)

Vol. 2 **Rheology of Materials and Engineering Structures** (Z. Sobotka)

Vol. 3 **An Introduction to Rheology** (H.A. Barnes, J.F. Hutton and K. Walters)

Vol. 4 **Rheological Phenomena in Focus** (D.V. Boger and K. Walters)

Vol. 5 **Rheology for Polymer Melt Processing**
(Edited by J-M. Piau and J-F. Agassant)

Vol. 6 **Fluid Mechanics of Viscoelasticity** (R.R. Huilgol and N. Phan-Thien)

Vol. 7 **Rheology: An Historical Perspective** (R.I. Tanner and K. Walters)

RHEOLOGY: AN HISTORICAL PERSPECTIVE

R.I. Tanner

P.N. Russell Professor of Mechanical Engineering, University of Sydney, Australia

and

K. Walters

Professor of Applied Mathematics, University of Wales, Aberystwyth, UK

1998
Elsevier
Amsterdam – Lausanne – New York – Oxford – Shannon – Singapore – Tokyo

ELSEVIER SCIENCE B.V.
Sara Burgerhartstraat 25
P.O. Box 211, 1000 AE Amsterdam, The Netherlands

Library of Congress Cataloging in Publication Data.
A catalog record from the Library of Congress has been applied for.

ISBN: 0 444 82945 8 (Hardbound)
ISBN: 0 444 82946 6 (Paperback)

Transferred to digital printing 2006
Printed and bound by CPI Antony Rowe, Eastbourne

Dedicated to our children

The Tanners
David, Jacqueline, Edwina, Ian and Rebecca

The Walters
Jeremy, Jonathan and Josephine

Preface

The science of Rheology remains a mystery to most people, even to some scientists. Some respectable dictionaries have been quite cavalier in their attitude to the science, the small Collins Gem dictionary, for example, being quite happy to inform us that a Rhea is a three-toed South American ostrich, whilst at the same time offering no definition of Rheology. We can argue long and hard about the reasons for this and similar oversights. Is it due to the fact that the science is interdisciplinary and doesn't fit well into any one of the historical scientific disciplines? Is it simply a result of being a relatively new science, with a relatively short life span?

An optimist would argue that there are now encouraging signs that rheology is being taken very seriously and it is certainly true that most of the larger modern dictionaries do attempt a definition. So we are likely to be spared such comments as "I recently came across a book on the history of rheology - but what is rheology?" This is obviously an encouragement to us.

We have both been rheologists for forty years or so and we have a great affection and enthusiasm for the subject. Hopefully, this will be evident to readers of the present book. It has been immensely enjoyable to write and the research has brought back many happy memories, since we have, of course, lived through some of the most exciting 'history' ourselves. As a general benefit, we believe the present book may be helpful to rheologists in locating older work of relevance and perhaps preventing the reinvention of discoveries already made.

At one time, we were plagued with the fear that most of the giants of the field were either dead or in retirement and we wondered whether the halcyon days of rheological research were now a thing of the past. Fortunately, in 1996, we both had occasion to spend several months at the Isaac Newton Institute in Cambridge. There we interacted with a good representative sample of a new generation of young rheologists and we were immensely encouraged. The field is certainly alive and well; it obviously has a bright future as well as a glorious past.

The selection of the work and personalities to be included in a living history will always please few; those left out will not be amused, and many of those included are likely to be displeased with their descriptions, relative to, say, those for Newton and Maxwell. We ask the reader's forgiveness in advance.

In a Preface, it would be expected of us to thank those who have assisted our endeavours and this we are very happy to do. The names are too numerous to record, but we do put on record our indebtedness to the scores who willingly supplied photographs, and, equally importantly, memories, insights and anecdotes. Thank you.

A significant amount of writing was done while we were both at the Isaac Newton

Institute in Cambridge. We wish to express our thanks to the council and staff of the Institute for the excellent facilities provided. One of us (KW) would also like to express his thanks to the Master and Fellows of Peterhouse, Cambridge, for the award of a Visiting Fellowship during 1996.

We wish to especially thank Mrs Pat Evans of the University of Wales, Aberystwyth, who has, with her usual understanding and expertise, typed and retyped the evolving text. Mr Aubrey Palmer has been responsible for the photographic work. Finally, we are happy to acknowledge the encouragement and support of Elizabeth and Mary.

Roger Tanner and Ken Walters
December 1997

Contents

Chapter 1

Introduction: The Ground is Prepared

1.1. In the Beginning

The science which we now know as rheology is in one sense very old; it certainly predates the formal introduction of the term in 1929 by centuries if not millennia. Scott Blair (1968, 1982) and Markovitz (1985), amongst others, have enthusiastically attempted to trace the origins by delving back into the histories of the Sumerian, Chinese, Egyptian, Indian, Greek and Jewish peoples. Water clocks and the concepts of "thinness" and "thickness" play an important role in the discussions and the names that figure most prominently are the Greek philosopher Heraclitus, the Jewish prophetess Deborah, and the Chinese philosopher Confucius (Kongfuzi). The philosophers Confucius (traditional dates 551-479BC) and Heraclitus (traditional dates 540-475BC) both thought about flow in philosophical contexts. Thus the famous dictum "panta rhei" (everything flows) attributed to Heraclitus has been taken as the motto of the (American) Society of Rheology. A more apt translation would be "everything is in a state of flux". In the Analects of Confucius ($\sim$ 250BC; see a new translation by Simon Leys, 1996) we find the statement in Section 9.17 "The Master stood by a river and said 'Everything flows like this, without ceasing, day and night"'. That two contemporaries came to a similar philosophical point is striking.

The famous song of the prophetess Deborah following the victory over the Philistines, recorded in the Book of Judges Chapter 5 v 5: "The mountains flowed before the Lord" is viewed as being of more immediate relevance to rheology, although, yet again, there is not universal acceptance that the biblical quotation stands up to critical rheological scrutiny. Some of the older translations have "melted" rather than "flowed", although the margin of the Authorized Version does contain the alternative translation of 'flowed'. Certainly, Marcus Reiner, who, as we shall see, figured prominently in the formal early developments of rheology, was convinced of the rheological connotation. In fact, he proposed the definition of the popular non-dimensional number known as the Deborah number on the strength of the reference, the idea being that everything flows if you wait long enough, even the mountains, (see §7.1).

1.2. Kinematics and Conservation Laws

Rheology, the science of deformation and flow of materials, can be viewed as dealing with three primary theoretical concepts: kinematics, conservation laws and constitutive relations. Kinematics is the science of motion, and it describes changes in the shape of bodies in time. Conservation laws deal with forces, stresses and the various energy interchanges arising from motion, while constitutive relations are special to classes of bodies (for example, viscous bodies) and serve to link motion and forces, thus completing the description of the deformation.

So far as kinematics is concerned, developments go back to antiquity: the law of the lever was known to Aristotle (4th century BC), and Archimedes (287-212BC) developed the concepts of the centre of gravity of bodies and of buoyancy in fluids. In the 14th century, we find the beginnings of the quantitative concepts of velocity and acceleration in the works of Albert of Saxony, the Oxford school, Nicole Oresme of Paris, and others. Oresme, who was Bishop of Lisieux in France from 1377 until his death in 1382, studied uniformly-accelerated motion. In the following centuries, kinematic concepts were further developed, so that, by the 16th century, these concepts were well understood, following work by Galileo and others. A terse history of early mechanics has been given by Dugas (1988).

The description of the kinematics of deformation and flow of materials arose later in the 18th century. In this connection, we note the work of the Bernoullis[1] and Euler[2] and the development of the theory of Elasticity in the early 19th century. By this time, the basic geometry of deformation and flow was completed. (For a relevant compendium of results see the text by Truesdell and Toupin, 1960).

Turning now to the conservation laws, it is clear that the conservation of *mass* is an idea of great antiquity. According to Rouse and Ince (1963), the first expression of the relation between cross-sectional area, velocity, volume and time occurred in the Dioptra of Hero of Alexandria (2nd century AD). The complete, modern statement of mass-conservation is due to Euler (1755), and it occurred in a series of articles in which he set up the equations of inviscid flow.

The exact conservation of *momentum* statement clearly cannot precede Newton (1687), in spite of earlier theories of impetus due to Buridan in mediaeval times (Dugas 1988). Without the concept of pressure, the application to continuum mechanics is in any case

[1]The Bernoulli family of Basel, Switzerland, was devoted to mechanics and mathematics. The most famous of the family are Johann Bernoulli (1667-1748), his elder brother Jakob (1654-1705) and Johann's son Daniel (1700-1782). The latter named and founded the science of hydrodynamics. He was born in Groningen in the Netherlands and educated in Basel, Switzerland. From 1725-1732, he held the Chair of Mathematics at the St Petersburg Academy. From 1732-1777, he was associated with the University of Basel, first as Professor of Anatomy and Botany and then as Professor of Natural Philosophy.

[2]Leonhard Euler (1707-1783) was a Swiss mathematician, born and educated in Basel, where he studied under Johann Bernoulli. He spent some time in St Petersburg with his great friend, Daniel Bernoulli (son of Johann). He also held appointments at Berlin as Director of the Berlin Academy of Sciences (1744-1766). He became Director of the Academy of Sciences in St Petersburg in 1766.

uninteresting. The recognition of the vector addition of forces by Simon Stevin[3] in 1586 (see, for example, Dugas 1988), was an essential prerequisite to Newton's work, and he also successfully attempted to define pressure. Pascal[4] (1663) eventually clarified the concept of pressure in a fluid at rest. Daniel Bernoulli's *"Hydrodynamica"* (1738) is important in this respect, since it applied the concept of pressure and something close to the principle of mechanical energy conservation to fluid motion. Many new ideas, including that of pressure tappings in the walls of ducts, appeared in Bernoulli's book, but the complete so-called Bernoulli theorem did not. Incredibly, five years later, in 1743, his father (Johann), then aged 76, published his own book on Hydraulics, and attempted to claim priority by using a false date of publication (1732). In this book, the modern form of the Bernoulli equation was presented; see Truesdell (1968).

Following the Bernoullis, d'Alembert (1752) investigated fluid flow in a rather cumbersome way, but he nevertheless wrote down the equations (in two dimensions) for inviscid fluids. Finally, in 1755, Euler set up the equations for inviscid fluid flow in a recognisably modern way, noting that pressure acted equally in all directions in a moving fluid; no question of viscous action was involved here, of course.

The modern *energy* conservation principle dates only from the mid-19th century and is due to Joule, Kelvin and others working in this period (see, for example, Smith and Wise 1989).

The history of the above two concepts of rheology, i.e. the description of deformation and flow and the enunciation of the balance laws, are really part of the great history of Continuum Mechanics, and we shall not dwell further on their developments in this book. One can refer for more details to Dugas (1988); the work of the Bernoullis (cf. D Bernoulli 1738) and of Euler (1755) must also be studied to gain an appreciation of the rapid developments that took place in the 18th century. (See also Love 1952 and Timoshenko, 1953 for readable accounts of some relevant topics in elasticity theory.) What is clear is that a thorough and clear appreciation of the basic geometrical concepts of deformation and flow and the important balance laws was already in place by the time rheological advances required them.

1.3. The Classical Extremes of Elasticity and Viscosity

The science of rheology, while it is clearly part of the whole field of continuum mechanics, is differentiated from the larger field by the need for special study of the many possible responses to deformation of diverse classes of materials. Therefore, we shall have a great deal to say about the recognition of the various classes of constitutive equations describing material behaviour; essentially this is what we understand as the quantitative study of rheology.

To facilitate the study, we shall need to look briefly at the important classical theories of elasticity and fluid flow, since these provide both the backdrop and the boundaries of

[3]Simon Stevin of Bruges (1548-1620) was a Flemish civil servant who eventually become responsible for major public works. Among many contributions to mechanics, he recognized and formulated the principle of the parallelogram of forces and the independence of hydrostatic pressure from the shape of the vessel. He also introduced decimal fractions into common usage.

[4]Blaise Pascal (1623-1662) was a French mathematician, physicist and Christian philosopher. Born at Clermont-Ferrand, he moved to Paris in 1631. In his honour, the SI unit for pressure is called the Pascal.

the modern science of rheology.

Our subject, as usual, begins in a qualitative way in antiquity, and we shall first sketch the early developments of the concepts of elasticity and viscosity.

Primitive ideas of the concept of *elasticity* go back at least to the development and use of the bow; in particular the use of the cross-bow goes back at least 3000 years (Needham 1959). However, the first published, quantitative concept of elasticity was that of **Robert Hooke**, whose law was published in 1676 as a Latin anagram (ceiiinossttuv = ut tensio sic vis) and in intelligible form in 1678.[5] This law emphasised the linear connection between force and extension in springs of all kinds.

Hooke had developed his law from experimental observations and he was very frustrated by the leakage of this intellectual property to rivals; he was interested in spring escapements for timekeeping in navigation and he had expected to benefit from inventions in this field.

Significant credit must clearly go to Hooke for his work on elasticity of solids and it is entirely proper that we should still speak of 'Hookean elastic solids'. However, we note that Robert Boyle, Hooke's onetime employer, had investigated the 'spring' of air and had published his results in the second edition (1662) of his 1660 work entitled *"New experiments physico-mechanical, touching the spring of air and its effects"*. He refers to "Mr Townelys hypothesis" (pV = constant) and proceeds to show that his experiments with a manometer tube demonstrated the truth of what should probably be called Townely's law. Hooke (1665) in his *"Micrographia"*, repeated these experiments and concluded (p225 of *Micrographia*) "... having lately heard of Mr Townely's hypothesis, I shap'd my course in such sort, as would be most convenient for the examination of that Hypothesis, the event of which you have in the latter part of the last Table... . From which Experiments, I think, we may safely conclude that the Elater of the Air is reciprocal to its extension, or at least very near". (Elater = pressure; extension = volume, here.) Later, in 1675, Edmé Mariotte rediscovered the law, and it is still known by his name in France. It represented the first constitutive law relating a stress to a state for fluids, and its discovery is remarkably close in time to Hooke's discovery of his spring law.

Nine years after the publication of Hooke's paper, **Isaac Newton** (1687) addressed the problem of steady shear flow in a fluid and the Principia contains the famous hypothesis: "the resistance which arises from the lack of slipperiness of the parts of the liquid, other things being equal, is proportional to the velocity with which the parts of the liquid are separated from one another". This 'lack of slipperiness' is what we now call viscosity.

The word 'viscosity' derives from the Greek and Latin words for mistletoe ($i\xi o\sigma$; viscum), a plant that produces a sticky substance from its berries. The concept of stickiness in fluid flow was appreciated in connection with the operation of the water clock (clepsydra) both in ancient Egypt and in China. The clepsydra seems to have originated in Babylon and migrated to Egypt; the oldest Egyptian clock yet found dates to around 1400BC (Cotterell and Kamminga 1990). Basically, since the operation of the clock depended on what is now called Poiseuille flow, it was found both in Egypt and in China that the water had to be heated in winter in order that time could be kept accurately (Needham 1959); we would now recognize that the governing feature here is the viscous

[5] A suitable translation would be "as the extension, so is the force." The 1676 anagram is an example of the common 17th century conceit of announcing a discovery by means of an anagram.

resistance in the capillary nozzle; the 'stickiness' of the water is important.

One can also quote from the Roman poet Lucretius (6-55BC). In his work *"Of the Nature of things"* (Latin: *De Rerum Natura*) he says

> "We see how quickly through the colander
> The wines will flow; on the other hand,
> The sluggish olive-oil delays; no doubt,
> Because 'tis wrought of elements more large
> Or else more crook'd and entangled".

One asks why it was not until the 19th century that such wisdom was incorporated into rheology?

Referring to Fig 1.1, we can write

$$\sigma_{yx} = \eta \frac{U}{d}, \tag{1.1}$$

where the shear stress is denoted by σ_{yx} and η is the coefficient of viscosity.

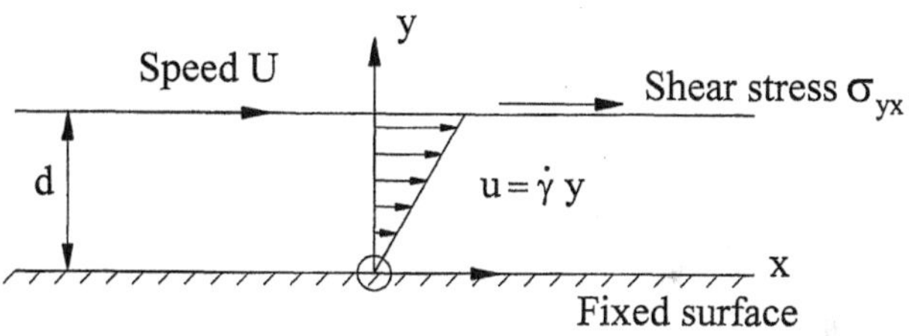

Fig 1.1. Simple shearing. Fluid is contained between two infinite parallel plates separated by a distance d. The top plate moves with a constant velocity U in the x direction. The shear rate $\dot{\gamma} = U/d$.

The formulation of this viscous law by Newton was not prompted by earthly consideration. He was anxious to discredit Descartes's theory of cosmic vortices, so he considered how a spinning planet could entrain fluid in a vortex motion. The space was assumed to be filled with a fluid, which obeyed a law such that the retarding force was proportional to the relative speeds of sliding of layers of fluid. Today, this seems like a mere guess for interstellar friction, but it has proved to be of the greatest utility for terrestrial purposes.

It appears unlikely that Newton ever carried out experiments to confirm (1.1) and in some respects he may not have taken it too seriously (see Markovitz 1968). It is certainly true that other scientists showed no interest in the hypothesis for at least a century. It is interesting, therefore, that, nowadays, Newton's postulate is held in very high regard by rheologists, who talk and write freely about Newtonian and non-Newtonian fluids.[6]

[6]Coleman et al (1966, p1) with good intentions, and sound reasoning, suggested that "Navier-Stokes' fluid" would be a more appropriate terminology to 'Newtonian fluid'. There has been no enthusiasm for the proposed change. The original terminology can be traced to Reiner (1929b).

The researches of Euler (1755) mark another stage in the development of the story. One finds in Euler's work, following pioneering studies by the Bernoullis, the use of partial differential equations to describe the dynamics of fluids and solids. The novelty here lies in the use of a small element of material and the consideration of the dynamics of the element. The resulting partial differential equations had then to be solved to find the motion.

Beam theory was initiated by Jakob Bernoulli and elaborated by Euler. No concept of a stress tensor embracing tension and shear was yet available. The beam theory involved tension and compression only, and, in the perfect fluids of d'Alembert[7] and Euler,the shear concept was unnecessary.

The perfect fluid concept of d'Alembert is perhaps the simplest constitutive equation of all. In modern notation, we can write it in the form:

$$\boldsymbol{\sigma} = -p\boldsymbol{I}, \tag{1.2}$$

where $\boldsymbol{\sigma}$ is the stress tensor, p is the pressure, and $\boldsymbol{I}$ is the unit tensor. Equation (1.2) implies that, in a perfect fluid, the total stress is just the negative of the pressure, with no shear stresses sustainable. The pressure has to be found as part of the solution of the problem in incompressible fluids and is not known in advance; it is partly determined by the boundary conditions.

The d'Alembert-Euler inviscid model (1.2) could not predict the observed pressure losses in duct flows and hence a more complex model involving shear stresses was needed. The importance of shear stresses, after Newton's beginning, seems to have been ignored until Coulomb[8], following Amontons (1699) and Leonardo da Vinci, investigated shear failure in soils and shear in beams (see, for example, Coulomb 1773, 1785). The failure criterion was subsequently refined by Tresca (1864, 1872).

Leonardo studied friction around 1470-1480, introducing the concept of the friction coefficient; one should perhaps call the Amontons/Coulomb friction laws after him. A detailed study of his work on friction is given by Dowson (1979); this book also reviews the early efforts to reduce friction by the use of lubricants and rolling bearings.

Coulomb (1801) also studied friction in fluids. Thus the importance of shear stress was well in evidence by the end of the 18th century, but no complete constitutive relation was yet at hand.

Thomas Young gave a set of lectures in 1807 in which he defined a modulus for extension. This was not what is now called 'Young's modulus', since the area of the rod was not involved, but it was a definite step towards the formulation of a constitutive relation. Young also recognized shear strain as a new concept.

Nowadays, we define Young's modulus through Navier's formula (published in 1826, see Timoshenko 1953):

$$\sigma_{xx} = E\epsilon, \tag{1.3}$$

[7] Jean le Rond d'Alembert (1717-1783) was a French mathematician and theoretical physicist. He studied law and was called to the Bar in 1738. Following a short flirtation with medical studies, he devoted the remainder of his life to mathematics. He was admitted as a member of the Academy of Sciences in 1741.

[8] Charles Coulomb (1736-1806) was a French physicist. He was born in Angoulême and educated in Mézières, graduating as a military engineer. He became President of the Institut de France in 1801.

where σ_{xx} is the tensile force per unit area of rod and ϵ is the elongation per unit length (see Fig 1.2(a)). We may also define, in a simple shear deformation,

$$\sigma_{yx} = G\gamma, \tag{1.4}$$

where γ is the shear strain (Fig 1.2b).

The work of Coulomb (1773) set the stage for the next advance, taken by Navier.[9] In May 1821, Navier read a paper to the French Academy, which was the first to give the fundamental equations of elasticity; it was not printed in extract until 1823 and in full until 1827 (Navier 1823, 1827a). The equations used were derived from a consideration of a system of spherical particles between which central (i.e. centre-to-centre) forces operated. The equations developed for the (small) displacement vector $\boldsymbol{u}$ were of the form

$$G(\nabla^2\boldsymbol{u} + 2\boldsymbol{\nabla}\Theta) + \boldsymbol{f} = \boldsymbol{0}, \tag{1.5}$$

where Θ is the volume strain of an element and $\boldsymbol{f}$ is the external body force vector per unit volume. G is the shear modulus, which is related to Young's modulus E and Poisson's ratio μ (Fig 1.2) by

$$E = 2G(1 + \mu). \tag{1.6}$$

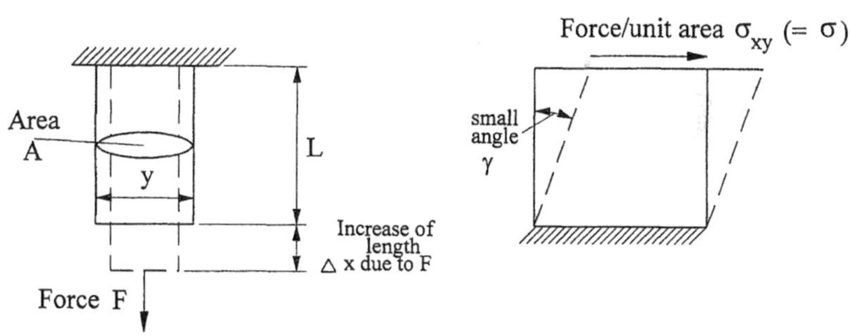

Fig 1.2 (a)left. Uniaxial extension. The stress σ_{xx} is given by $\sigma_{xx} = F/A$ and the strain ϵ by $\Delta x/L$. There is also a diminution of width Δy. The original width is y, and the ratio $(\Delta y/y)/\epsilon$ is called Poisson's ratio μ.
(b)right. Shear deformation. The shear modulus G is defined by $\sigma_{yx} = G\gamma$.

In fact, in Navier's particle theory, Poisson's ratio had the value 0.25 and, accordingly, only one elastic constant was needed to describe the materials. Navier's memoir is remarkable in that it produced nearly correct macroscopic results from a microstructural theory.

[9]Navier (1785-1836) was trained as an engineer at the Ecole Polytechnique in Paris and then went to the Ecole des Ponts et Chaussées, where he taught after 1818.

Very soon afterwards, the (symmetric) stress tensor concept was introduced by **Cauchy** (1827) in France, and this was used, for example, by Poisson[10] (1829) in deriving solutions to elasticity problems. The necessary studies of motion and deformation were also set in place by Cauchy (1827, 1828, 1829), and the basic equations of classical elasticity became available for the first time:

$$\boldsymbol{\sigma} = \lambda \Theta \boldsymbol{I} + 2G\boldsymbol{e}, \tag{1.7}$$

which is an expression of Hooke's law for isotropic bodies. Here λ and G are the so called Lamé constants and $\boldsymbol{e}$ is the infinitesimal strain tensor given by

$$\boldsymbol{e} = \frac{1}{2}\left(\boldsymbol{\nabla u} + \boldsymbol{\nabla u}^T\right), \tag{1.8}$$

$\boldsymbol{\nabla}$ being the gradient operator. Θ is the dilatation, or volume strain, given by $\boldsymbol{\nabla}.\boldsymbol{u}$. In Navier's theory, $\lambda = G$, a special case of equation (1.7), giving the Poisson ratio the value 0.25.

Thus, by 1827, a complete, realistic, constitutive equation, which is still in widespread use, was available for classical (small-strain) elasticity.

Navier (1821, 1822, 1827b) was also the first to address seriously the motion of viscous fluids. He again looked at a molecular model, and, making the assumption that the forces between the molecules were diminished by an amount proportional to the velocity of approach, Navier derived the equations:

$$\boldsymbol{f} - \boldsymbol{\nabla} p = \rho \frac{D\boldsymbol{v}}{Dt} - \eta \nabla^2 \boldsymbol{v}, \tag{1.9}$$

where $\boldsymbol{v}$ is the velocity vector, ρ is the density, D/DT is the material derivative, $\boldsymbol{f}$ is a bodyforce per unit volume and η is a viscosity to which Navier attached no special significance.

Stokes[11] (1845, 1851) realized that the resistance to attempted or actual volume change is an essential feature of fluid motion, and he wrote

$$\boldsymbol{\sigma} = -p\boldsymbol{I} + \Lambda \theta \boldsymbol{I} + 2\eta \boldsymbol{d}, \tag{1.10}$$

where the pressure p is a function which can now be identified with the thermodynamic pressure in a fluid at rest, this being a function of density and temperature. θ is the rate of change of volume, given by $\boldsymbol{\nabla}.\boldsymbol{v}$. Λ is a second viscosity, while η is the shear viscosity. The tensor $\boldsymbol{d}$ is the rate-of-deformation tensor:

$$\boldsymbol{d} = \frac{1}{2}(\boldsymbol{\nabla v} + \boldsymbol{\nabla v}^T). \tag{1.11}$$

While there were measurements of η by 1845, there were no measurements of Λ. (In most cases this is still true; for compressible fluids undergoing rapid deformation, the

[10] Siméon Denis Poisson (1781-1840), born at Pithivièrs, Loiret in France, is mainly celebrated for his contributions to electricity and magnetism, but he also made substantial contributions to the theory of elasticity.

[11] George Gabriel Stokes (1819-1903) was a British physicist. He was born in Screen, Sligo, in Ireland and educated in Dublin, Bristol and Cambridge. He was Lucasian Professor of Mathematics at Cambridge from 1849 until his death in 1903. Stokes was President of the Royal Society from 1889-1890.

response is more complex than the model (1.10) would suggest). Hence, Stokes made an hypothesis, still often used, that

$$\Lambda = -\frac{2}{3}\eta, \tag{1.12}$$

which ensures that pure volumetric changes, without shearing, do not dissipate energy. For the case of an *incompressible* fluid, the term $\Lambda\theta\boldsymbol{I}$ can be combined with the pressure term, to yield

$$\boldsymbol{\sigma} = -p\boldsymbol{I} + 2\eta\boldsymbol{d}, \tag{1.13}$$

where the pressure p is not now determined by thermodynamic behaviour, but has to be found as part of the solution of the boundary-value problem. We shall mainly deal with incompressible fluids in our sequel, and the no-volume-change constraint then leads to the extra condition:

$$\boldsymbol{\nabla}.\boldsymbol{v} = 0, \tag{1.14}$$

which is the conservation of mass equation for the incompressible case.

The experimental verification of the classical theories of elasticity and fluid flow was somewhat delayed. In the case of elasticity, experimental difficulties (see Love 1952) meant that it was not until the 1850's that reasonably sound evidence emerged against the restriction $\mu = 0.25$ associated with the early theories.

In the case of fluid flow, the picture was totally confused because most experimental work on pipe flow dealt with turbulent flow, which was not recognized until Reynolds's (1883a,b) work. Stokes (1845) attempted to compare his theory of pipe flow with the experimental work of Bossut (1797) and du Buat (1816), but without success.

The later works of Poiseuille[12] (1846) and Hagen (1839) did show that the flow rate was proportional to the pressure drop and to the fourth power of the capillary diameter and would have conformed to Stokes's theoretical calculations. However, the actual formula for the flow rate was not published by Stokes, although he did publish the correct velocity field. It was left to Wiedemann (1856) and Hagenbach (1860) to produce the well known formula for the volumetric discharge rate Q from a capillary of radius R:

$$Q = \frac{\pi R^4}{8\eta}(\Delta p/L). \tag{1.15}$$

Here, L is the length of the capillary, Δp is the pressure drop and η is the (constant) viscosity.

While the above discussion appears to support the Stokes analysis, it should be remembered that the agreement between experiment and the theory embodied in (1.15) is conditional upon the use of the 'no-slip' boundary condition at the wall of the capillary, which was employed by Stokes. Later in the 19th century, various investigators took this

[12]Jean Léonard Marie Poiseuille, French physiologist, was born in Paris in 1799. He received his MD in 1828 and practised medicine in Paris. In 1842, Poiseuille was elected to the Académie de Médicine in Paris. He died in 1869. Poiseuille has given his name to the unit of viscosity in the CGS system (the Poise).

point seriously (see Dryden et al 1950) and came to the general conclusion that fluids did indeed 'stick' to solid walls, thereby confirming Stokes's treatment. It now appears that this conclusion is only guaranteed for fluids made up of small molecules (like air, water and oils) at low shear stresses, and we shall need to return to this point later.

We have now essentially sketched the development of the mother lodes of rheology - classical elasticity and fluid dynamics. Rheology goes beyond these special cases and, by common consent, excludes them. To proceed, we now need to review the deviations from the classical situations that were noticed in the 19th century.

1.4. Non-Classical Behaviour

Even before the firm verification of the classical theories of elasticity and fluid dynamics was completed, marked deviations began to be observed. The first puzzle came in the form of deviations from Hooke's law. Perhaps this should not have been too surprising, since hysterisis and loss of springiness had been noticed in bow technology for millennia. The ability to deform metals permanently was also a matter of common observation; such behaviour, which is clearly not elastic, is a part of the general study of 'plasticity', the history of which is covered by Hill (1950) and Love (1952).

For our present purpose, it is convenient to begin by highlighting the work of Wilhelm Weber (1835, 1841) on silk threads. At that time, there was a general interest in improving galvanometer construction and the use of silk fibres for instrument suspensions was common. Weber noticed that the elasticity of silk fibres in tension was not perfect. He applied a tensile load to a silk fibre and noticed an immediate (elastic) extension; this was followed by a continued slow extension with time. Removal of the load led to an immediate contraction. This kind of behaviour was already well known in metals, but, to Weber's surprise, it was found that the silk fibre eventually recovered its original length. This differentiates the elastic-after-effect (elastische nachwirkung) from previously observed elastic, plastic or creep behaviour in metals and marks a distinct point of departure for our *viscoelastic* studies.

While Weber only conducted what we now call creep and recovery tests, he also stated what would happen in a relaxation test, where a sudden strain was imposed and maintained. Weber also discovered the need to 'condition' the silk fibre by imposing a few load/unload cycles, before reliable, repeatable behaviour was observed. The mechanical conditioning eliminated some non-recoverable creep observed in the first few loading cycles.

In the second paper, Weber (1841) fitted the observed extension $x(t)$ with a law of the form:

$$\dot{x} = bx^m. \qquad (1.16)$$

He found $m = 6.82$, and was able to solve (1.16) and fit his experiments reasonably well; the response curve was not exponential.

Weber's experiments clearly demonstrated the essence and complexity of rheological behaviour. The fact that the fibre eventually returned to its original length meant that he was essentially dealing with what we would now call a viscoelastic solid. There were clear differences from the plastic behaviour in metals observed by Coriolis (1830) in lead specimens, which would nowadays be referred to as creep. Hence, Weber's observations are the first serious study of the class of elastico-viscous deformations which have so intrigued rheologists.

More work on elastic after-effects was carried out by Wertheim (1847) and Kupffer (1857). Following Weber's observations, Wertheim discovered after-strain in human tissues, possibly marking the beginning of biorheology. He was also active in measuring Poisson's ratio, showing that it was not equal to 0.25 universally and he measured wave speeds in bars to support his conclusions (Wertheim, 1851). Previously, he had had a "polemic" with St Venant over the theory of torsion.

Kupffer (1857) clearly recognized the torsional damping of metals and, in 1865, Thomson (later Lord Kelvin)[13] used the term 'viscosity of metals'.

The work of the Kohlrausches, father and son, is worthy of note. In 1847, R Kohlrausch, in Berlin, was interested in electrometers and performed work on the torsional response of fibres; this work was reported in a footnote to a long paper on electrometers.

In 1863, Friedrich Kohlrausch[14], after his father's death, carried out more work on the torsional response of a number of materials, including rubber and glass fibres, with some curve fitting similar to Weber's early work on extension. He fitted some of his retraction curves as a power law $x = At^{\propto}$. He also investigated the decrement of an oscillating system with a 'lossy' fibre and he established the essential linearity of the torsional phenomenon, separating the time and magnitude effects of the response, which Weber had not done. Later, Kohlrausch (1866) clarified the difference between elongation and torsion, and studied the effect of temperature on material properties. In 1876, he extended the ideas to bending, by which time he knew of Boltzmann's (1874) theoretical work. F Neesen (1874) introduced multiple exponential terms of the type now familiar to describe his after-effect experiments.

We now need to look at the state of *theoretical* work at these times. We have already noted Kelvin's (Thomson 1865) concept of 'viscosity of solids'. In 1867/68, **James Clerk Maxwell** put forward the idea that the "viscosity in all bodies may be described independently of hypothesis" by the equation:

$$\frac{d\sigma}{dt} = E\frac{d\epsilon}{dt} - \frac{\sigma}{\lambda}, \tag{1.17}$$

where σ is the stress and ϵ the strain. E is Young's modulus and λ is a time constant. No real explanation was given for this equation and Maxwell used it to calculate gas viscosity, given by the product $E\lambda$. He was aware of the experiments of Weber and Kohlrausch and noted that the simple exponential decay of stress implied by (1.17) did not agree with their data; he proposed that λ should depend on the stress, an idea which has since been

[13]W Thomson (Lord Kelvin 1824-1907), was a British physicist. Born in Belfast, he was educated at home by his father (a Professor of Mathematics). In 1834, he entered the University of Glasgow and in 1841, the University of Cambridge. In 1846 be became Professor of National Philosophy in Glasgow, where he created the first physics laboratory in a British University. He was President of the Royal Society from 1890-1894 and was granted a peerage in 1892, taking the title 'Baron Kelvin of Largs'. He is buried next to Isaac Newton in Westminster Abbey, London.

[14]Friedrich Wilhelm Kohlrausch (1840-1910), son of R Kohlrausch, was born in Rinteln-on-Weser, Germany. He is chiefly famous for his work on the electrical conductivity of solutions. He studied under Wilhelm Weber at Erlangen (PhD 1863).

frequently revisited. It is perhaps ironic that the concepts of the rivals Hooke and Newton were united forever by Maxwell in his equation (1.17).

The next theoretical work that we are aware of is that of Oskar E Meyer (1874) of Breslau (now Wroclaw in Poland). He assumed that the shear stress σ and strain γ could be written in the form:

$$\sigma = G\gamma + \eta\frac{d\gamma}{dt}, \tag{1.18}$$

where G and η are material constants. This expression actually describes what is now known as the Kelvin-Voigt body. Kelvin (see Thomson 1865, 1878) did experiments on the damping of metals and applied the concept implied by (1.18) without writing down any formulae; Voigt (1889, 1892) generalized the Meyer idea to anisotropic media. In justice, one ought to refer to a Kelvin-Meyer body, or simply a Meyer material.

An important contribution to the subject was made by **Ludwig Boltzmann**, who is best known for his work on the kinetic theory of matter and the concept of entropy. His three rheological papers, published in 1874, 1877 and 1878, were to have a significant impact on the mathematical theory of viscoelasticity.

Boltzmann's early work in 1874, written as he attained his thirtieth year, was apparently motivated by the lack of generality in Meyer's (1874) formulation, and the paper contains a long criticism of Meyer's work.

Considering the isotropic viscoelastic case, Boltzmann assumed that the stress at time t not only depended on the strains at that time, but also on those in previous times; it was explicitly assumed that the longer the interval from the present to the past time, the smaller would be the contribution to the stress resulting from a given strain. This is an expression of the now familiar principle of 'fading memory'. The assumption of (linear) superposition was also made, with a footnote that states that the principle of superposition will not hold for large ("starke") deformations. This was then followed immediately by the now accepted general theory of viscoelasticity, in the form:

$$\sigma_{xx}(t) = \lambda\Theta(t) + 2Ge_{xx}(t) - \int_0^\infty d\omega\Phi(\omega)\Theta(t-\omega) - 2\int_0^\infty d\omega\phi(\omega)e_{xx}(t-\omega). \tag{1.19}$$

Here, Φ and ϕ are memory functions and $\omega = t - \tau$, τ being a past time. The shear components corresponding to (1.19) were also given in the usual form; only ϕ is relevant in a shear deformation, since the terms containing Θ are unnecessary in this case. For these components, Boltzmann then deduced several ways of finding $\phi(\omega)$ from torsional experiments using various strain patterns; relaxation, free vibration and short steps of strain were among the patterns considered.

Some experiments on glass fibres in torsional vibration were shown to agree well with Boltzmann's calculations and the closing remarks in the paper concern the way the general theory collapses to the viscous case for certain forms of the memory function ϕ.

Boltzmann's (1874) paper clearly deserves its reputation as a classic - it criticised an existing theory (Meyer's), it gave a new formulation and tested it against new experiments on glass fibres; it finally closed with some far-sighted predictive remarks.

Meyer (1878) in turn attacked Boltzmann vigorously on all fronts, arguing that his theory was not based on an atomic hypothesis and violated ideas then current about atoms; he was also unhappy about the new experiments.

We have already indicated that Meyer's (1874) theory was nothing more than that for the so called Kelvin-Voigt model, which was to be published later by Kelvin (1878) and Voigt (1889, 1892).

It is evident that, although both the Maxwell model (1.17) and the Meyer model (1.18) are valid expressions of two kinds of materials showing a viscoelastic response, they are inferior in generality to Boltzmann's equation (1.19). Maxwell recognized this fact and used the Boltzmann formulation in his Encyclopaedia Britannica article of 1878. In fact, the Maxwell model is a special case of the Boltzmann theory for shearing, with

$$\phi(s) = \frac{\eta}{\lambda^2}\exp(-s/\lambda); \; G = \eta/\lambda, \tag{1.20}$$

where η is a viscosity, λ is a time constant and G is an elastic modulus. (From now on, λ will generally denote a time constant, as opposed to a modulus as in 1.7 and 1.19.)

The Kelvin-Voigt-Meyer model is best regarded as a Boltzmann model with zero ϕ and the addition of a viscous component, unless one is prepared to admit generalized functions for ϕ.

The complete Boltzmann theory was fully three-dimensional and is the general form of linear viscoelasticity. We may quote and agree with Markovitz (1977) who concluded that it was the "first successful theory of rheology".

Boltzmann's two later papers (1877 and 1878) were replies to criticisms by Meyer and others and essentially added nothing fundamental to the subject, although they did contain carefully worked examples.

1.5. Appraisal

A widely held view marks rheology as a relatively modern science, which has really come into prominence in the latter half of the 20th century. This we concede, but the present chapter has shown that its origins are in antiquity and that the 19th century in particular contains much research of relevance. This was carried out by a number of famous scientists and they provided a solid foundation for the mushrooming activity of the 20th century.

The advances in previous centuries were not monotonic and, like in so many other fields, there were the inevitable personality clashes. The antipathy between Hooke and Newton and between Meyer and Boltzmann are prime examples of this.

What is clear is that, by the turn of the 20th century, there was a general acknowledgement of the existence of materials that could not be classified by either of the classical extremes of Hookean elastic solids and Newtonian viscous fluids.

1.6. Robert Hooke

Robert Hooke was born on the Isle of Wight in 1635. He had a great natural aptitude for making mechanical toys, and he was initially educated at home by his father, a clergyman. He went to Westminster school at the age of 13 and then to Oxford University, where he met Robert Boyle, the chemist, who was a council member of the newly-formed Royal Society. He assisted Boyle with some of his experiments and in 1660 he was hired as Curator of Experiments by the Royal Society at the urging of Boyle. He struggled financially and in addition to the curatorship and a lectureship, he was also Professor of Geometry at Gresham College in London.

During his time at the Royal Society, Hooke investigated an enormous range of subjects, including mechanics, light, chemistry, philosophy, botany and microscopy. His *Micrographia* (1665) contains a description of a new compound microscope and discussions of the physics of thin plates and views on combustion, besides impressive drawings of what he saw in the microscope. This book alone would have made his reputation as a first-rate scientist.

Hooke's *Micrographia* contains comments of immediate relevance to today's experimental rheologist: "I have often thought that probably there might be a way found out to make an artificial glutinous composition, much resembling, if not fully as good, nay better, than that excrement, or whatever substance it may be out of which the silk worm wire-draws his clew... . This hint... may give some ingenious person an occasion of making some trials, which if successful... I suppose he will have no occasion to be displeased."

Hooke was clearly a very versatile and an exceptionally energetic man. He assisted with the surveying and architecture of the rebuilt London, following the Great Fire of 1666 (cf. Cooper, 1997) . He also did much to transform the Royal Society from an amateur's club to a professional body.

To some, Hooke was sociable and respected, 'a person of great suavity and goodness'. To others he was too cynical and miserly to be much liked. In his later years he was described as being 'of middling stature, something crooked, pale faced, and his face but little below, but his head is lardge; his eie full and popping, and not quick; a gray eie. He haz a delicate head of haire browne, and of an excellent moist curle' (cf. Keynes 1960). He was nevertheless combative and this led to many disputes on priority, notably with Newton, who was seven years younger than Hooke. Newton and Hooke were always at loggerheads and it seems likely that Hooke's status as a scientist was deliberately lessened by Newton, following Hooke's death in 1703. A sympathetic biography is that by Espinasse (1956). No portrait of Hooke is known to have survived. In its place we reproduce a famous picture (Fig 1.3) from his work on springiness (*De potentia restitutiva* 1678), which is his lasting contribution to rheology.

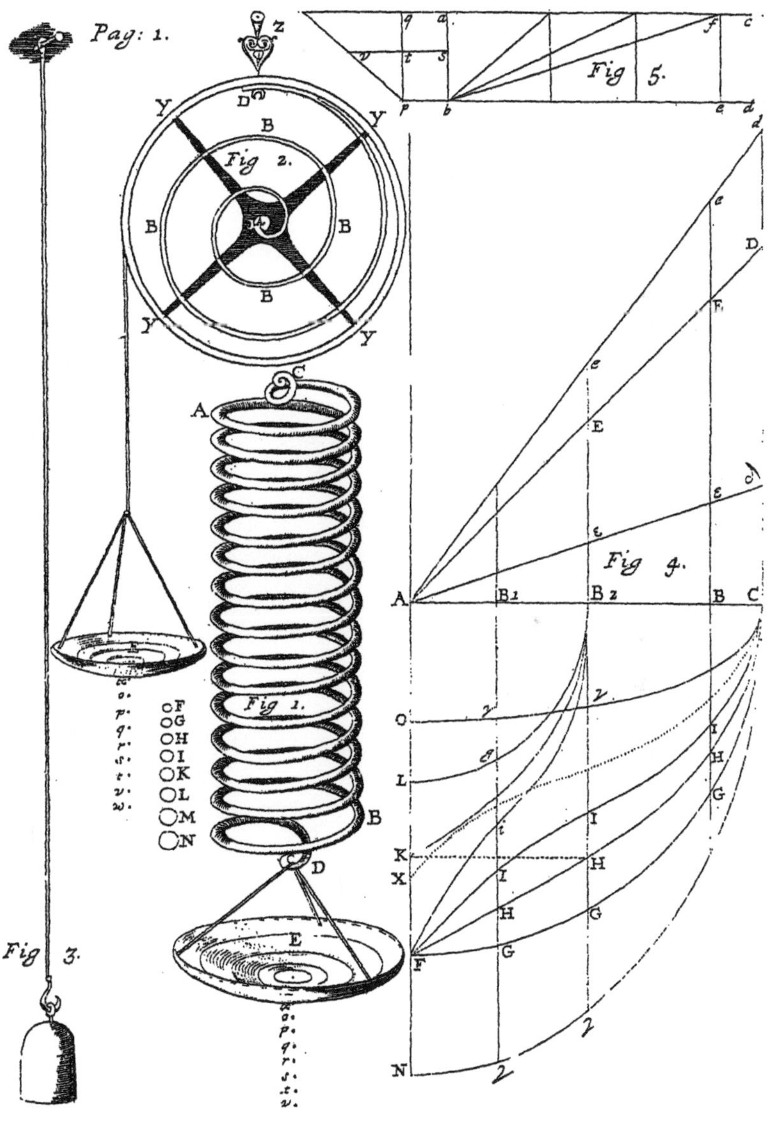

Fig 1.3. Hooke's spiral springs, from *De potentia restitutiva*, 1678.

1.7. Isaac Newton

SIR ISAAC NEWTON
1642—1727
"Newton was the first to formulate a hypothesis regarding the magnitude of the force required to overcome viscous resistance and to treat a case of motion in a viscous fluid." Cf. *Principia*, Lib. ii, Sect. IX. Emil Hatschek

This portrait of Newton and the accompanying text is taken from the opening pages of the first issue of the first volume of the *Journal of Rheology* published in 1929.

Isaac Newton was born in Woolsthorpe, Lincolnshire, on Christmas day 1642, according to the Julian calendar then in force in England; for the rest of Europe, which was already using the modern Gregorian calendar, his date of birth was 4th January 1643. His father, an illiterate but reasonably prosperous farmer, died three months before the birth. Three years later, his strong-minded mother remarried and moved to another village, leaving Isaac Newton at Woolsthorpe, to be brought up by a grandmother. His stepfather, a wealthy clergyman, died eight years later, and Newton's mother then returned to Woolsthorpe. Some have seen this separation from his mother as an important factor in the shaping of the suspicious and neurotic personality of the adult Newton (see, for example, Gjertsen 1988).

Newton was educated at Grantham Grammar School and then from 1661 at Trinity College, Cambridge, becoming a Fellow in 1667. He made many remarkable discoveries in mathematics at an early age. During the time he was absent from Cambridge in 1666 due to the plague, he began his work on gravitational problems and the motion of the moon. It was not until 1684 that Edmond Halley interested him in writing up his work. Eventually, it became known as the *Principia*; it was published in 1687.

The *Principia* contains Newton's famous hypothesis about the response of fluids to a steady shearing motion, which assured his place in our History.

Between 1692 and 1694, Newton was unwell, suffering from insomnia and nervous trouble. In 1695, he was offered and accepted the Wardenship of the Mint, a post he retained until his death in 1727; he was buried in Westminster Abbey, London. An exceptional biography of Newton is that of Westfall (1980), who discusses not only his scientific work, but also his philosophical and theological studies, as well as his work as Master of the Mint.

Aside from problems in his relations with Robert Hooke, which we have already alluded to in our discussion of Hooke, Newton was also in priority disputes with Leibnitz over the Calculus. As we have hinted, these difficulties with people were probably there from boyhood (Westfall 1980) and cannot be overlooked, any more than can his devotion to alchemy and religious studies. The picture is of a self-sufficient, secretive personality, mellowing somewhat in old age. He never married and, given these facts, it is hard not to have some sympathy with his opponents. He remains a remote, mysterious person, one of the supreme intelligences of all time.

1.8. Augustin-Louis Cauchy

Cauchy as a young academician. Half-length portrait by Boilly 1821 (see Belhoste 1991).

Augustin-Louis Cauchy was born in Paris in 1789, the year of the French Revolution. His father was a successful royalist government lawyer and he had to leave Paris at the beginning of the Revolution for the village of Arceuil. Laplace and the chemist, Berthollet, also lived in Arceuil and, during the time of Napoleon, young Cauchy came to know these famous scientists. Lagrange, when visiting Laplace's house, noticed Cauchy's mathematical ability. In 1816, Cauchy became a member of the Academie and held three Professorships in Paris. In 1830, he left these positions, because he declined to swear allegiance to Louis Philippe; he was extremely pious and a strong Catholic. Cauchy was forced into exile and went to Turin. He finally returned to the Ecole Polytechnique in 1838; he died in 1857.

Cauchy's interest in elasticity was generated by Navier's memoir of 1821, which was presented to the Academie. In 1822, he formulated the general theory of stress, and in so doing defined the stress tensor and derived the equations of motion. These have been the cornerstone for much of theoretical rheology ever since and Truesdell in his *"Essays in the history of mechanics"* asserts that "Clearly this work of Cauchy marks one of the great turning points of mechanics and mathematical physics" (Truesdell 1968, p188).

He published seven books and about 800 papers, and has around 16 fundamental concepts and theories to his credit; certainly he was one of the greatest mathematicians in history.

Several authors have written about him; we note the works of Valson (1868), Bell (1953) and Belhoste (1991).

Claude-Alphonse Valson (1826-1900) was Professor of Calculus at the Université de Grenoble and his volume on Cauchy devotes nearly as much space to Cauchy's work for the Catholic Church as it does to his mathematics. It is uncritical and adulatory, and it is no surprise to learn that Valson also espoused Cauchy's views on society and religion. Valson praised Cauchy's teaching, but the evidence shows he was a classroom failure (Belhoste, 1991).

The well-known work of Bell has a chapter on Cauchy and it is also adulatory, but concentrates on mathematics. The more recent study of Cauchy by Belhoste (1991) is more balanced and one has to admit that Cauchy's narrow views, plus his many mean-spirited actions and disputations over priority with his mathematical contemporaries, do not make him an attractive personality, despite his extraordinary contributions to mathematics and mechanics.

1.9. James Clerk Maxwell

James Clerk Maxwell was born in Edinburgh, Scotland in 1831. After what to some was an unpromising beginning, he suddenly blossomed and his first scientific paper was published by the Royal Society of Edinburgh when he was only thirteen. He entered Edinburgh University at the age of sixteen.

Maxwell went up to Trinity College, Cambridge, in 1850 to embark on a distinguished scientific career. In 1856 he moved back to Scotland as Professor of Natural Philosophy at Marischal College, Aberdeen, the move being motivated at least in part by a desire to be near his ailing father, his mother having died when Maxwell was only nine. He married the daughter of the Principal of Marischal College.

In 1860, Maxwell moved to Kings College, London, and remained there until the death of his father in 1865. There followed a lengthy stay at his family home in Scotland, where he combined the life style of a gentleman farmer with that of a research scientist.

In 1874 Maxwell was persuaded to return to Cambridge as the first Cavendish Professor of Experimental Physics. There he set up the famous Cavendish Laboratory, which was to become a unique institution, headed by a succession of men of genius.

Maxwell is universally accepted as the greatest theoretical physicist of the 19th century and he is sometimes spoken of as the 'Father of modern physics'. He certainly had an immense influence on subsequent generations of scientists. Albert Einstein, for example, concluded that "one scientific epoch ended and another began with James Clerk Maxwell" and R A Millikan, another Nobel Laureate, wrote of 'one of the most penetrating intellects of all time'.

It is evident that materials science was peripheral to Maxwell's major scientific interests. However, his contributions of immediate relevance to rheologists have been unusually influential. Best known is his work on "The Dynamic Theory of Gases" (Maxwell 1867, 1868) in which he introduced the famous linear differential equation relating stress and strain; this allowed for the first time the idea of a relaxation process in a viscous fluid.

Less well known is his experimental work on the determination of the "thickdom" or viscosity of gases and his perceptive distinction between 'solids' and 'liquids'. This is highlighted in Markovitz's historical essay on *"The emergence of rheology"*: "In his writing on states of matter, Maxwell drew a careful distinction between solids and liquids and, in many of his publications, pointed out that materials such as pitch are fluids: If therefore, we define a fluid as a substance which cannot remain in permanent equilibrium under a stress not equal in all directions, we must call these substances (cold pitch and asphalt) fluids, though they are so viscous that we can walk on them without leaving any footprints. What is required to alter the form of a soft solid is a sufficient force, and this, when applied, produces its effect at once. In the case of a viscous fluid, it is *time* which is required, and if enough time is given, the very smallest force will produce a sensible effect" (Markovitz 1968).

Maxwell interacted with Lord Kelvin, who was also to leave his mark on early developments in rheology. An example of the association is contained in a letter written by Maxwell in 1865 to a relative, C H Cay (see, for example, Campbell and Garnet, 1884), the point at issue being the interpretation of experimental data from a viscometer. The letter was written from Glenlair, the family house in the Scottish Highlands; it contains a hint of Maxwell's quaint and suggestive turn of phrase: "I set Professor W Thomson a prop. which I had been working with for a long time. He sent me 18 pages of letter of suggestions about it, none of which would work; but on Jan 3rd, in the railway from Largs, he got the way of it, which is all right; so we are jolly, having stormed the citadel, when we only hoped to sap it by approximations".

By common consent, Maxwell was somewhat diffident, a man of quiet disposition, unassuming, devoid of any trace of pomposity, but of great warmth and sure character. He had a strong Christian faith and his obituary in *Nature* paid tribute to this important aspect of his life: "His simple Christian faith gave him a peace too deep to be ruffled by bodily pain or external circumstances". In 1879, Maxwell died of cancer at the relatively early age of 48.

1.10. Ludwig Boltzmann

Ludwig Eduard Boltzmann was one of the greatest theoretical physicists, famed for his work on gas kinetic theory, heat and entropy. The son of a taxation official, he was born on the evening of Mardi Gras, (February 20th, 1844) just before Ash Wednesday, and he later attributed (in jest) his rapid changes in moods, from high to low, to this aspect of his beginning. He received his doctorate from the University of Vienna in 1866, with Josef Stefan as supervisor. He worked on the kinetic theory of gases for his PhD, a subject to which he would later contribute much more original work. After two years, at the early age of 25, he was appointed Professor of Mathematical Physics at the University of Graz. He moved to Munich in 1891, then back to Vienna in 1894, when a Chair became vacant on the death of Stefan. Ernst Mach was appointed as a Professor of Philosophy at the University of Vienna in 1895, but Boltzmann was at philosophical odds with Mach, and so he moved on to Leipzig in 1900. Mach's early retirement in 1901 finally left Boltzmann able to reclaim his own chair at Vienna, since it had not yet been filled.

In his career, Boltzman defended the atomistic viewpoint vigorously. It is perhaps surprising that debate on these issues should be taking place at such a late date, but the question of the 'reality' of atoms and molecules was a very serious point of contention in the late 19th century, and controversy was only laid to rest in the early 20th century. Mach was one of the opponents of Boltzmann, together with the chemist Wilhelm Ostwald. Despite a victory for atomicism at a public meeting in 1895, the attack on it continued, and Boltzmann began to feel insecure about the outcome. His health declined and he became depressed, possibly about the future of his work and ideas. However, the work of

Rayleigh, Ramsay, Einstein, Smoluchowski, Becquerel and the Curies supporting atomic theories were all published in Boltzmann's lifetime, so it is not clear that one can attribute his severe depressions to this cause alone; probably his general state of health, including failing sight, was a sufficient reason for his ultimate depression. On September 20th, 1906 at the age of 62 years, he committed suicide by hanging, while staying at the beautiful Bay of Duino (near Trieste) with his wife and daughter. Soon after his suicide, his views were vindicated by the work of Perrin and others and the acceptance of the atomic viewpoint was inevitable.

Boltzmann was short and stout, played the piano well, wrote humorous poems and loved nature and the countryside. He was an excellent teacher and gave very well-attended popular lectures; there is a whole volume of Populäre Schriften available. He was said to be kind towards students and never to have failed anyone taking his courses in his later years.

His most admired work was on the kinetic theory of gases and on thermodynamics; the celebrated H-theorem relating the entropy (S) to the logarithm of the number of available states (W) is certainly his most important and widely applicable result; the formula $S = k \log W$ is engraved on his tombstone in Vienna. He also gave a theoretical proof of Stefan's fourth-power radiation law for a black body. The 1926 Encyclopaedia Britannica entry does not mention his work on viscoelasticity, nor does the 1983 Broda biography, but the three papers written in 1874, 1877 and 1878 are his fundamental contribution to our subject. Despite their importance to rheology, they appear almost as a continuum aside in a life devoted to atomic and molecular dynamics.

Chapter 2

The Growing Years Before 1945

2.1. The Beginning of Experimental Fluid Rheology

Much of the previous chapter has dealt with problems arising from *solid* rheology, in particular from fibre mechanics. We now need to address the beginnings of serious experimental *fluid* rheology. Some of the relevant background in Newtonian flow has already been discussed in the previous chapter, but it is important for us to highlight the innovative experimental work of Hagen (1839), a civil engineer, and Poiseuille (1835, 1846), a physician, both of whom studied flow in small-diameter tubes for different reasons; Hagen was looking for basic hydraulic information and Poiseuille was interested in blood flow. The experimental technique of Poiseuille was exemplary, and is still worthy of study. The same can be said of the seminal work of Couette[1]

The work of Hagen, Poiseuille and Couette set the stage for the 20th century preoccupation with *non*-Newtonian liquids. At this time, non-classical behaviour was most clearly observed through the variation of viscosity with shear rate. The first experiments that unequivocally demonstrated a non-Newtonian viscosity varying with shear rate seem to be those of Theodore Schwedoff, who was Dean of Sciences in Odessa in the late 1880s. He used a thin-gap Couette-type apparatus, and was therefore able to infer shear stress and shear rate fairly accurately. Two of his papers appeared in 1889 and 1890 in the Journal de Physique. In the first, he described concepts of viscosity and springiness that he sought to measure; he also dealt with fluids that showed a yield point, such as weak gelatine gels. He found that the material relaxed after a step strain, and he tried to use Maxwell's ideas to describe what he saw. In his second paper, Schwedoff (1890) concentrated on viscosity. He concluded: "(the viscosity) is not constant, as one usually supposes: it varies with the speed of shearing". Since he was dealing with gel materials that had a yield stress,

[1]Maurice Marie Alfred Couette was born in 1858 in Tours, France, the only son of a cloth merchant. In 1877, he obtained a bachelor's degree in mathematics from the Faculty of Sciences in Poitiers. He then enrolled in the newly opened Free Faculty of Sciences in Angers and, in 1879, he received a degree in physics. After completing one year's voluntary military service, Couette settled in Paris. From 1887, he studied at the Physical Research Laboratory at the Sorbonne, completing his doctoral thesis on "Studies in liquid friction" in 1889. In 1890, at the invitation of the Chancellor, Couette returned to the Catholic University of Angers, where he was to spend the next 43 years as Professor of Physical Sciences. He gave up university teaching in 1933 and died ten years later. A full account of Couette's life and scientific achievements is provided by Piau et al (1994).

he also saw that the viscosity tended to an infinite value as the shear rate went to zero. He was able to show a constant viscosity for glycerine, but using a 1% gelatine solution, there was a seven-fold variation in viscosity as the shear rate varied by about 15:1.

To describe the results, Schwedoff invented the generalized Maxwell model for the shear stress σ :

$$\frac{d\sigma}{dt} = G\dot{\gamma} - \frac{1}{\lambda}(\sigma - \sigma_y). \tag{2.1}$$

Here σ_y is a shear yield stress. We note that J G Butcher (1877) invented the terms "elastico-viscous" and "elastico-plastico-viscous", of which (2.1) is an example, to describe material behaviour; the latter term is, happily, now rare.

The 1890 paper of Schwedoff was the forerunner to a multitude of papers on variable viscosity effects in a plethora of materials which were to occupy much of the literature of the first half of the 20th century (e.g. Hatschek 1913). According to Scott Blair (1938), there was a tendency to label all anomalous behaviour as manifestations of 'plasticity', with no clear idea as to what that meant.

In 1922, Bingham published an important book entitled "Fluidity and Plasticity". It contained a mammoth 82 page bibliography and provided a great deal of information on measurements for various systems, including gel-like materials with a yield stress. It also included an important discussion on 'wall-slip' in viscometers. Strangely, the book does not have a very lengthy discussion on variable-viscosity systems, which probably reflects the mood at the time of writing.

This mood had to change, and it did so as a result of rheometrical experiments on a multitude of different materials. So, by 1939, Scott Blair, in the March issue of 'British Plastics', was able to report that "Rheological methods are in practice applied to an enormous variety of materials. Among the more obvious applications are the following: metals (elastic and plastic properties), oils, paints and varnishes, bitumens, tars and pitch, gums, inks and starches, flours for bread and biscuit making, cheese, butter, cream and milk, tooth pastes, soaps and toilet creams, ceramic clays, dyestuffs and fibres of all kinds..." (see also Scott Blair 1938). In order to classify the plethora of possibilities in a simple way, the British Rheologists Club published a working chart on rheological behaviour (Nature **149**, 1942, p702) and invited further comments. Figure 2.1 contains a response from L Bilmes (1942) in the form of a circular chart.

References to many of the more relevant contributions to the expanding literature are provided in the books by Houwink (1937), Scott Blair (1938) and Philippoff[2] (1942). The latter is a remarkable book, first published by Steinkopff in Germany during the Second World War and reprinted two years later at Ann Arbor in the United States. The book is an important source of reference on the state of colloidal rheology in the early 1940s. We note that the often misunderstood term 'thixotropy', which we discuss in Chapter 7,

[2]Wladimir Philippoff was born in Peterhof, Russia in 1907 and moved to Germany in 1924, receiving his Dr Eng (Elec Eng) degree in 1934. From 1932-45 he worked at the Kaiser Wilhelm Institute of Chemistry on dynamic testing in viscoelasticity. After working for the Military Government for three years he moved to the USA in 1948. In 1959, he joined the Esso Research and Engineering Center in Linden, N. J. where he worked until his retirement. He was awarded the Bingham medal in 1962.

was invented in this period.

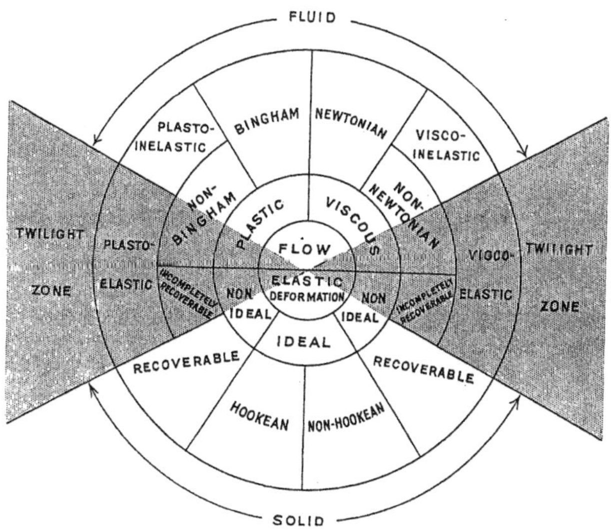

Fig 2.1 The Rheological Chart proposed by Bilmes (1942).

With the increasing interest in fluid rheology came the need to describe nonlinear fluids in an appropriate manner and a large number of empirical models describing the observed shear stress-shear rate behaviour were invented. This was essentially a curve-fitting exercise, and, to be attractive, the models had to contain as few constants as possible.

Most notable amongst the growing list of available models were the Bingham (1922) model, the Ostwald (1925)-de Waele (1923) power-law model, the Herschel-Bulkley (1926) model and the Ellis and Williamson models (see Houwink 1937, Tanner 1988). These are still important today in giving a compact, manageable description of shearing flow. One could say that if variable viscosity and linear viscoelasticity were the sum total of rheology, then it had all been done by 1930.

The empirical shear stress-shear rate models were used to derive the relevant flow rate-pressure drop relationships for capillary flow, and these could then be compared with experiment. However, it was soon realized that a procedure was needed to solve the important *inverse* problem, i.e. given a set of flow rates Q and pressure drops Δp from a given capillary, what shear stress σ - shear rate $\dot{\gamma}$ curve does this imply (if one exists)? This question was successfully addressed by **Weissenberg**, the provenance of the solution being unexpected, since, at the time (1928), Weissenberg was working at the Berlin crystallographic laboratory of R O Herzog at the Kaiser-Wilhelm Institut für Faserstoffchemie Berlin-Dahlem, which was not primarily interested in fluid mechanics. Weissenberg's interest in these matters can be traced to a paper with Herzog (1928) entitled "On the thermal, mechanical and x-ray analysis of swelling" which was published in Kolloid-Zeitschrift. It is a remarkably wide-ranging paper and sets out a programme of research on the relations of material and molecular structures to thermal and mechanical properties. Whilst this was clearly only a preliminary statement of a vast programme of work, it was recognized that material behaviour might be described as an interplay of kinetic, thermal and elastic energy, and a triangular diagram with vertices labelled with these three forms of energy was displayed. In a surprising switch from this very general scene to a particular problem, on their p281, the first general analysis of the inverse capillary flow begins.

In this analysis, it is assumed that the velocity gradient (dw/dr) is a function of the shear stress, i.e.

$$\frac{dw}{dr} = f(\sigma). \tag{2.2}$$

Then follows their equation (2), which states that the volume discharge rate Q, the capillary radius R, length L and pressure drop Δp are related by

$$\frac{Q}{R^3} = F\left(\frac{R\Delta p}{2L}\right) = F(\sigma), \tag{2.3}$$

where

$$f(\sigma) = \frac{1}{2\pi}\left[\frac{d^2F}{d(\ell n\sigma)^2} + \frac{5dF}{d\ell n\sigma} + 3F\right]. \tag{2.4}$$

We are unable to see how this clearly incorrect formula was derived.

The *correct* inversion formula was published twice in 1929. The first publication, submitted on March 14, 1929 was authored by Eisenschitz, Rabinowitsch and Weissenberg (1929), but before it appeared Rabinowitsch (1929) published again (August 1929) in the Zeitschrift für Physikalische Chemie, a more accessible journal. However, both Rabinowitsch, in his 1929 paper, and Eisenschitz, in a 1933 paper, acknowledge unequivocally that the inversion method was due to Weissenberg. This brilliant analysis will bear repeating.

If $w(r)$ is the axial velocity distribution with respect to a cylindrical polar coordinate system, we have

$$Q = 2\pi\int_0^R w(r)\, r\, dr. \tag{2.5}$$

Recognizing that the fully-developed wall shear stress σ_w is given by

$$\sigma_w = \frac{R\Delta p}{2L}, \tag{2.6}$$

(2.5) can be written, after an integration by parts,

$$Q = 2\pi\left[\frac{wr^2}{2}\right]_0^R - 2\pi\int_0^R \frac{r^2}{2}\frac{dw}{dr}dr. \tag{2.7}$$

A crucial step in the Weissenberg analysis is the assumption of no slip at the wall, so that

$$w(R) = 0; \tag{2.8}$$

in which case the first term on the right hand side of (2.7) vanishes. (Later, Mooney (1931) generalized the analysis to allow for slip.)

Assuming the no-slip condition and introducing the 'fluidity function' $\phi(\sigma)$, which is the inverse of the viscosity and is given by

$$\frac{dw}{dr} = -\sigma\phi(\sigma), \tag{2.9}$$

where $\sigma = r\Delta p/2L$,a positive quantity, (2.7) becomes

$$Q = 8\pi\left(\frac{L}{\Delta p}\right)^3 \int_0^{\sigma_w} \sigma^3\phi(\sigma)d\sigma. \tag{2.10}$$

If we replace Δp by $2\sigma_w L/R$ and define q as $Q/\pi R^3$, we find

$$q = \frac{1}{\sigma_w^3}\int_0^{\sigma_w} \sigma^3\phi(\sigma)d\sigma.$$

Differentiation with respect to σ_w gives the desired result; using (2.9), we have

$$\left|\frac{dw}{dr}\right|_w = \dot{\gamma}_w = q(\sigma_w)\left[3 + \frac{d(\log q)}{d(\log \sigma_w)}\right], \tag{2.11}$$

so that the $\dot{\gamma}_w(\sigma_w)$ curve can be constructed. This formula has been the basis for all reliable capillary viscosity measurements since 1929.

The extension to plane (slit) flows follows immediately; a factor 2 appears on the right hand side of (2.11) and suitable redefinitions of q and σ_w have to be made. Many further inversion formulae for other viscometric flows are available (see Coleman et al 1966, Walters 1975 and Tanner 1988).

The correct application of formulae like (2.11) requires a knowledge of the pressure drop along the capillary in fully-developed flow; i.e. away from the inevitable flow disturbances at the ends of the capillary. This was (and is) not always available and convenient empirical means have been devised to overcome the problem. The first was proposed by Couette and has been called the Couette method. This involves using two capillaries of the same diameter but different lengths L_1 and L_2, the flow condition being the same in both. The pressure loss in a fully developed region over a distance $L_1 - L_2$ is then calculated by subtraction (see Piau et al 1994).

A modification of the Couette method was proposed much later by Bagley (1957) and the use of the so called 'Bagley correction' has become standard practice in much modern capillary and slit rheometry.

At this point, we need to return to the work of Weissenberg. Following his contributions to crystallography, with the invention of the Goniometer in 1924, (see, for example, Harris 1973), we see his attention turning to the general problem of material behaviour. The capturing of the angles of x-ray diffraction no doubt sharpened his appreciation of three-dimensional states, with great advantage to rheology.

We have already referred to Weissenberg's inversion formula, which appeared in his initial foray into the field in 1929. A year later, he co-authored a paper which described an oscillating viscometer (see Harris 1973). Two long papers followed, one written in German (Weissenberg 1931) and one written in French, while he was in Southampton, UK (Weissenberg 1934). These discuss the application of tensor methods to material problems. In his 1931 paper, the idea that the extra stress tensor is determined by the flow history is expressed. The impact of these long papers was less than it might have been. Nevertheless the stirring of an interest in three-dimensional rheology can be seen. Weissenberg published no more work until 1947 and we shall discuss this in Chapters 6 and 7.

2.2. Linear Viscoelasticity

Linear viscoelasticity is a subject which stretches in its influence and importance from the early days of rheology to the present day. Certainly, the work of Maxwell, Boltzmann, Voigt, Kelvin and others, already discussed in Chapter 1, falls within the restricted area which is now commonly referred to as 'linear viscoelasticity'. It remains an important area of research and most modern characterization studies involving viscoelastic materials are likely to involve the measurement and interpretation of data arising out of appropriate experiments, be they creep, stress relaxation or small amplitude shear flow experiments. Indeed, one could say that the bulk of the fundamental theoretical research in linear viscoelasticity was completed by the close of the 19th century.

The early years of the 20th century witnessed the introduction of 'mechanical models' and these have proved to be a popular means of characterizing linear-viscoelastic behaviour. In these models, Hookean elastic deformation is represented by a spring and Newtonian flow by a dashpot. Sometimes, a 'slider' is introduced to represent a yield criterion (see, for example, Reiner 1945a). The mechanical-model analogy involves the association of force, extension and time in the models with stress, strain and time in the material.

The Maxwell model, described mathematically by the differential equation (1.17), is represented by a spring and dashpot in series, whereas the Kelvin/Voigt model described mathematically by equation (1.18) requires the spring and dashpot to be connected in parallel. The characterization of the behaviour of more complex viscoelastic behaviour is brought about by the introduction of a series of Maxwell or Kelvin/Voigt elements connected in parallel or series, respectively. These more complex models were independently suggested by J J Thomson (1888) and Wiechert (1893).

To some, the introduction of the spring/dashpot analogy was an unnecessary and unwanted incursion into a well-developed field. Typical of this attitude is: "While Maxwell left it for later generations to obscure his straightforward notions of continuum mechanics by the intermediary of springs and dashpots, his approach to the subject may fairly be said to have dominated it until 1945" (Truesdell 1960b).

Others have appreciated the didactic role of the spring-dashpot representation (cf Reiner 1945a, Tschoegl 1989) seeing it as a way of studying viscoelastic response without the need to go into mathematical detail, and the comments of Ferry (1970) would be typical of many exponents of mechanical models. "When a material exhibits linear viscoelastic behaviour, its mechanical properties can be duplicated by a model consisting of some suitable combination of springs... and dashpots (pistons moving in oil). Two such models are illustrated in Figs 1.1 and 1.2. To simulate a real material, the model may require an infinite number of units with different spring constants and flow constants..." (see our Fig 2.2).

"It is of course possible to develop the subject of viscoelasticity without recourse to mechanical models, but they are introduced here for convenience in visualizing the combination of Hookean solidlike and Newtonian liquidlike characteristics... . The model represents only the macroscopic behaviour and does not necessarily provide any insight into the molecular basis of viscoelastic phenomena; its elements should not be thought of as corresponding directly to any molecular processes".

The first reference to mechanical models is in the text of Poynting and Thomson (1902) and a full descriptive account is in the 'First Report' by Burgers (1935). Figure 2.3 taken from this report illustrates a four-element model, which is sometimes referred to as the 'Burgers model'. It contains a number of influential simpler models as special cases. So, when the spring α_2 and dashpot ϕ_2 are absent, we have the 'Maxwell model', while, when α_1 and ϕ_1 are absent, we have the 'Kelvin/Voigt model'. The model with α_1 missing is known as the 'standard linear solid'.[3]

The three-element model for an elastico-viscous liquid, obtained by omitting the spring α_1, has played an influential role in the development of the subject. (It would be more correct to say that the *differential equation* corresponding to this model and usually written[4]

$$\boldsymbol{T} + \lambda_1 \frac{\partial \boldsymbol{T}}{\partial t} = 2\eta_0 \left[1 + \lambda_2 \frac{\partial}{\partial t} \right] \boldsymbol{d} \tag{2.12}$$

has played an influential role in the development of the subject.) For example, Jeffreys (1929) applied the equation to interpret problems associated with the crust of the earth. Later, Fröhlich and Sack (1946) showed that, within the linearity constraint, a theoretical analysis for a very dilute suspension of elastic solid spheres in a viscous liquid resulted in an equation like (2.12), and Oldroyd (1953) did the same for a dilute emulsion, the suspended particles in this case being liquid droplets, with the interfacial tension providing the restoring force; as a result, the individual droplets resist changes of shape, thus imparting viscoelasticity to the emulsion. To accommodate 'interfacial slipping', Oldroyd showed that equation (2.12) has to be replaced by

$$\boldsymbol{T} + \lambda_1 \frac{\partial \boldsymbol{T}}{\partial t} + \nu_1 \frac{\partial^2 \boldsymbol{T}}{\partial t^2} = 2\eta_0 \left[1 + \lambda_2 \frac{\partial}{\partial t} + \nu_2 \frac{\partial^2}{\partial t^2} \right] \boldsymbol{d}. \tag{2.13}$$

By implication, the mechanical model for such a system is more complicated than the one shown in Fig 2.3.

---

[3]The standard linear solid has also been referrred to as the 'Boltzmann solid' (Davies 1950), but this terminology is not now extant.

[4]$\boldsymbol{T}$ is the extra stress tensor, the total stress $\boldsymbol{\sigma}$ being given by $\boldsymbol{\sigma} = -p\boldsymbol{I} + \boldsymbol{T}$.

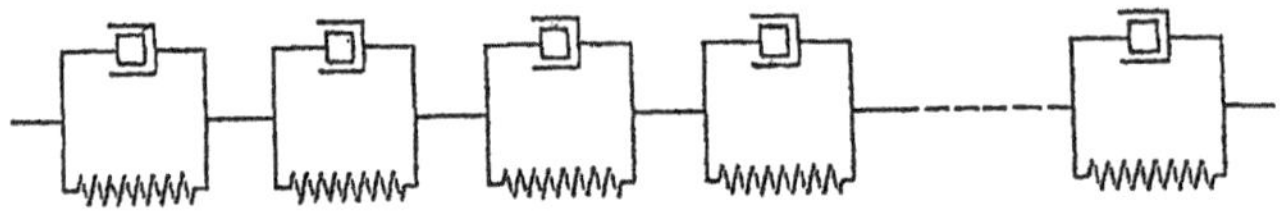

FIG. 1-1. Generalized Voigt model of springs and dashpots representing the mechanical behavior of a viscoelastic material.

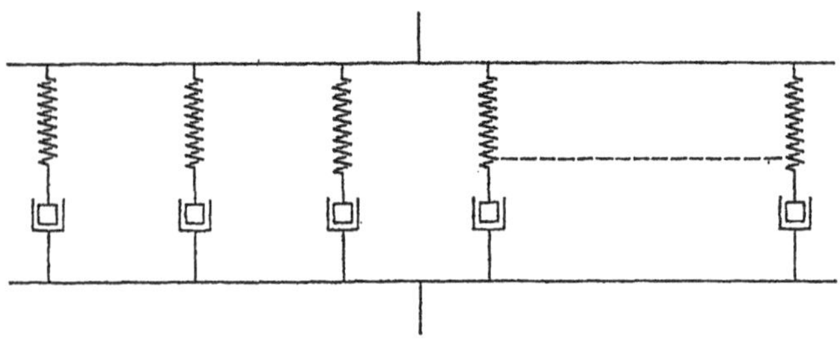

FIG. 1-2. Generalized Maxwell model of springs and dashpots representing the mechanical behavior of a viscoelastic material.

Fig 2.2. The canonical forms from Ferry (1970).

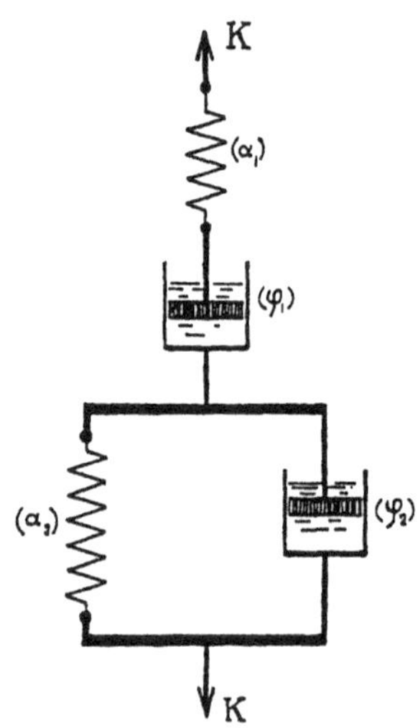

Fig 2.3. The four-element mechanical model reproduced from Burgers (1935).

It is obviously possible to envisage more complex mechanical models than those already discussed, but Roscoe (1950) showed that all models, irrespective of their complexity, could be reduced to two canonical forms, which are usually taken as the generalized Kelvin/Voigt model and the generalized Maxwell model (see Fig 2.2). Roscoe also showed that a saving in the number of elements often results from the use of canonical forms and he gave rules for deducing the number of elements in the canonical form from the arrangement of elements in any complex model.

By a suitable choice of the model parameters, the canonical forms themselves can be shown to be mechanically equivalent and Alfrey (1945) gave methods for computing the parameters of one canonical form from those of the other. In the same paper, he showed how the differential equation relating stress and strain could be obtained quite generally from either of the canonical forms and vice versa.

The mechanical-model representation of linear viscoelastic behaviour results in at best an enumerable infinity of relaxation times, but the extension to a continuous distribution of relaxation times is accomplished with mathematical ease. Numerous influential texts in the post Second World War years discuss the procedures involved, most notably Gross (1953), Staverman and Schwarzl (1956), Alfrey and Gurnee (1956) and Ferry (1970).

The resulting equations are often written in the form:

$$\boldsymbol{T} = \int_{-\infty}^{t} \phi(t-t')\boldsymbol{d}(t')dt', \tag{2.14}$$

where

$$\phi(t-t') = \int_{0}^{\infty} H(\tau) \exp\left[-(t-t')/\tau\right] d\tau, \tag{2.15}$$

although other equivalent forms of the distribution function $H(\tau)$ are sometimes used.

A careful comparison of (2.14) with (1.19) clearly demonstrates that all these developments were anticipated in Boltzmann's (1874) work, which was quite general.

Solution of boundary value problems for linear viscoelastic materials, including temperature variations via the time-temperature superposition principle (see Chapter 6), made great strides in the post Second World War years.

The time-temperature superposition idea was extended to moving media by Morland and Lee (1960). This key paper introduced the idea of a local "material clock time", which varies at a given particle as that particle encounters variable temperatures during its kinematic history; it still remains the basis of many calculations and has been extended to the nonlinear case. Further extensions of the variable clock time to include pressure and stress effects have also been considered. Further descriptions of these principles are given by Tanner (1988) and Huilgol and Phan-Thien (1996).

One may refer to the book of Bland (1960) for a discussion of the correspondence principles linking the solution of linear viscoelastic problems to 'corresponding' linear elastic solutions.

2.3. The No-Slip Boundary Condition

In Newtonian fluid mechanics, it is now widely accepted that the no-slip boundary condition holds; i.e. particles of fluid adjacent to the wall move with the wall velocity.

However, the history of this concept is not as straightforward as it might seem and rheologists have, for good reasons, been more concerned about the validity of the concept than workers in Newtonian fluid mechanics.

Interest in the subject goes back at least to Coulomb (1801), who carried out some viscosity-measuring experiments using an oscillating disc in water. He tried coating the disc with tallow and also sprinkling on sandstone, neither of which made any differences to the observations. He therefore concluded that no slip was occurring and that a true fluid property was measured. Meyer (1861) confirmed these results.

Poiseuille's seminal (1835, 1846) work actually implied the no slip boundary condition, since eventually Hagenbach (1860) and Wiedemann (1856) were able to show that there was concordance with Stokes's (1845) solution for laminar flow, which used the no slip condition. Helmholtz (see Bingham 1922) did allow for slip, with the result that, for a Newtonian fluid, the flow rate/pressure drop relationship took the form:

$$Q = \frac{\pi}{8}\left(\frac{\Delta p}{L\eta}\right)\left[R^4 + 4\bar{\lambda}R^3\right], \tag{2.16}$$

where $\bar{\lambda}$ is a slip coefficient.

Several experimenters thought that they had found slip and Helmholtz was able to compute values of $\bar{\lambda}$ from various sources. Some thought that mercury should slip at a glass wall, because of its non-wetting nature. Warburg (1870) investigated this case and found no slip; his results were later confirmed.

Whetham (1890) carried out a very extensive set of experiments with various surfaces, as did Couette between 1888 and 1890 (see Piau et al 1994); Couette also investigated slip in the turbulent regime. In all cases, the conclusion was that, within experimental error, no slip was seen.

Interestingly, Bingham (1922, p29-35) devoted considerable space to the problem of slip and concluded: "These results seem to make it quite certain that, whether the liquid wets the solid or not, there is no measurable difference between the velocity of the solid and the liquid in contact with it, so long as the flow is linear".

The last phrase is intriguing and presumably meant that Bingham had some doubts about the non-Newtonian case; but for Newtonian liquids he saw no slip, and this view has been accepted in fluid mechanics and has even extended to non-rarefied gases.

The situation in *non*-Newtonian fluid mechanics is more complex, and, by 1929, rheologists were actively considering slip, especially for rubbery materials. Mooney (1931) extended Weissenberg's capillary-flow relation to include slip, since he knew that rubbers did not adhere fully to solid walls.

Much later, Oldroyd (1949) appealed to 'slip at the wall' in an attempt to explain Toms's (1949) turbulent drag reduction results. Similarly, in plasticity theory (see Hill 1950), slip between metal and die wall is commonplace. Pearson and Petrie (1965) devised a stability theory of polymer flow which depended on wall slip. Further, there were numerous papers showing jumps in discharge rates in capillaries (see, for example Vinogradov and Malkin 1980). Thus the question of slipping (or not) continued to arise.

It is now convenient to survey the various possibilities concerning slip.

(i) In solutions, it is clear that the presence of a solid wall may alter the concentration of solute near the wall, thereby inducing a low-viscosity layer and apparent slip. In this case, the solvent does not slip at the wall. This possibility is discussed in detail in the recent survey of Barnes (1995).

(ii) In a molten polymer, the wall will alter the configuration space available to molecules near the wall, even if slip does not occur, thereby altering the material behaviour next to the wall.

(iii) Slip may occur, separately from or in addition to (ii) above.

Effect (i) is relatively well understood and need not be discussed further; in blood flow it has a name - the Fåhraeus-Lindqvist effect (1931; see also Goldsmith and Mason 1967).

Effect (ii) must certainly occur, but so far only simulations can detect it and the altered layer seems generally not to penetrate far into the bulk material.

In connection with effect (iii), Pearson and Petrie (1968) made a basic contribution to the subject, although the work was not widely noticed, and the subject languished until Ramamurthy (1986) showed, unequivocally, that real slip can occur in extrusion. Hatzakiriakos and Dealy (1991) then followed with a set of revealing experiments using a novel sliding viscometer.

Pearson[5] and Petrie (1968) consider the length scales involved near the wall, nominating L as the scale of the apparatus (e.g. a diameter), ℓ as the scale of the roughness on the surface of the apparatus, and ℓ_p as the scale of the microstructural entities making up the fluid. It is the relative scales of these lengths, they said, that determines, to a large extent, the boundary conditions. In the various cases:

1. $L >> \ell >> \ell_p$ is the typical condition for small-molecule fluids (water, light oils). In this case the small scale of the asperities means that any flow over them is at very low Reynolds number, and viscous forces dominate near the boundary. Here, the argument of Richardson (1971) shows that actual adhesion to the wall is not necessary in order to see, on the gross scale (L), an apparent no-slip condition, even if slip occurs on the ℓ_p scale.

2. $L >> \ell_p >> \ell$. This condition is typical of coarse powders in smooth containers, where slip at the wall is often seen.

3. $L >> \ell \approx \ell_p$. This is often the case for large molecule fluids near smooth walls, and the question of slip may well depend on chemical and/or mechanical adhesion, in contrast to the cases above.

In summary, with the benefit of decades of both theoretical and experimental interest, we can now conclude that at least three factors are of importance:

[5] John Richard Anthony Pearson was born on September 18th, 1930. His early schooling took place in Egypt and then England. After two years military service, he entered Trinity College, Cambridge in 1950 and he obtained a BA in the Mathematical and Mechanical Sciences in 1953. There followed a year at Harvard University, where he obtained an AM. He returned to Trinity in 1954 and was awarded a PhD in 1957. His postdoctoral career has involved industrial appointments with ICI and Schlumberger and academic appointments in Chemical Engineering at Cambridge and Imperial College, London.

Anthony Pearson was awarded a Silver Medal by the Plastics Institute in 1964, an ScD by the University of Cambridge in 1975, and a Gold Medal by the British Society of Rheology in 1986. In 1980, he was elected a Foreign Associate of the US National Academy of Engineering.

(i) The Pearson/Petrie scaling argument means that no effective slip can occur when molecular size is smaller than the wall roughness scale. This really takes care of all the classical evidence on liquids (and non-rarefied gases) and shows that a no-slip condition is appropriate in these cases; chemical adhesion is not needed to avoid slip; see also recent work by Sarkar and Prosperetti (1996).

(ii) For large molecules (relative to the wall roughness scale), the temperature and adherence (chemical) properties may be of great significance in setting the critical shear stress at which slipping occurs.

(iii) Normal pressure may assist in reducing slip.

We have chosen to locate our discussion of slip within the context of "The growing years before 1945", but it is clear that interest in the subject stretches beyond that time frame. Indeed, its study is still in vogue.

2.4. Theoretical Non-Linear Developments 1880-1945

In one sense the present chapter has already addressed a number of topics of a theoretical nature, mostly in the linear regime. The inverse problem for capillary flow and the general question of slip clearly involve a significant theoretical component. But we now want to concentrate on the distinct theoretical problems involved in 'constitutive equations'. Developments in the period 1880-1945 were sparse but significant, and they have not received the attention they merit.

We begin with a few remarks about the generalization of classical infinitesimal-strain elasticity theory to include *finite* strains. The slow developments in this area can be traced via the works of Truesdell (1952), Truesdell and Noll (1965) and Murnaghan (1937). The effective solutions to some key problems was achieved by Rivlin(1948b); see also Rivlin (1959) and the books by Treloar (1958) and Ogden (1984). A noteworthy constitutive model for rubber was introduced at this time by Mooney (1940); it is now often referred to as the Mooney-Rivlin equation.

The early experiments of Poynting (1913) showed that a rubber rod lengthened when it was twisted, against all expectations of small-strain theory. This clue was not followed up until the work of Rivlin and Saunders on torsion was published in 1950.

The theory of *plasticity* also made great strides following work by Hencky, von Mises, Geiringer and Prager. Hill (1950) provides a convenient history of developments in this area.

We shall now consider developments in non-Newtonian fluid mechanics in more detail.

At the turn of the present century, formulations for relaxing bodies were considered and Ladislaus Natanson of the University of Cracow wrote (see Natanson 1901, 1903)

$$\frac{D\boldsymbol{\sigma}}{Dt} = \Lambda(tr\boldsymbol{d})\boldsymbol{I} + 2G\boldsymbol{d} - \frac{1}{\lambda}(\boldsymbol{\sigma} + p\boldsymbol{I}), \tag{2.17}$$

where $D/Dt(= \partial/\partial t + \boldsymbol{v}.\boldsymbol{\nabla})$ is the usual material derivative; Λ and G are moduli, λ is a relaxation time, p is the pressure and $\boldsymbol{\sigma}$ is the stress tensor. Stanislaus Zaremba (1903), also from Cracow, disagreed with Natanson over the formulation, primarily because (2.17) was not written in a frame rotating *and* translating with the medium, which he thought was proper. To rectify this omission, Zaremba introduced the feature of a corotating derivative, replacing D/Dt by $\mathcal{D}/Dt$,[6] where

$$\frac{\mathcal{D}\sigma_{ij}}{\mathcal{D}t} = \frac{D\sigma_{ij}}{Dt} + \sigma_{im}\omega_{mj} + \sigma_{mj}\omega_{im}, \tag{2.18}$$

ω_{ji} being the vorticity tensor, defined in this case as

$$\omega_{ji} = \frac{\partial v_i}{\partial x_j} - \frac{\partial v_j}{\partial x_i}, \tag{2.19}$$

where v_i is the velocity vector. Zaremba also replaced p/λ in (2.17) by $(tr\boldsymbol{\sigma})/\lambda_n$, where λ_n is another relaxation time.

Although Natanson seems to have been Zaremba's academic superior, the latter wrote (1903) "Il est donc prouvé que les equations de M Natanson sont absoluments inexactes". As one might expect, several more defending and attacking papers were published in 1903 in the Bulletin of the Academy of Sciences of Cracow, but there were few followers and little external interest.

This series of papers terminated in 1903, and of course Zaremba's point is a vital one; the idea of having to consider the rates of change of tensor quantities carefully, and the subsequent writing down of an objective constitutive equation (with what is now called a corotational stress derivative) is due to Zaremba.

Jaumann (1905, 1911) also considered these corotational derivatives and, despite Zaremba's priority, the corotational rates are often termed 'Jaumann derivatives'. This is not to imply that Jaumann's work was generally appreciated or acknowledged and, as late as 1961, Prager wrote "Jaumann's work does not seem to be well known", and commented further that the definition of the corotational derivative was often used in the literature without reference to Jaumann. He singled out Fromm (1933), Zaremba (1937), Thomas (1955), Noll (1955) and Hill (1959). Except for Zaremba, whose 1903 paper clearly predates that of Jaumann, Prager's remark is just.

The next scientist to consider Zaremba's work seriously seems to have been Heinrich Hencky.[7] Hencky is best known for his work on plasticity (1923, 1924), but he also

[6]Throughout this book, we shall use standard tensor notation, wherever necessary. Covariant suffices are written below and contravariant suffices above, and the usual summation convention for repeated suffices is assumed; usually we simply use Cartesian tensors.

[7]Information about Heinrich Hencky's life is not easy to find; we have not discovered an obituary. He received his Diploma in Civil Engineering in the Technical University of Munich in 1908 and a Doctorate from Darmstadt in 1913. He spent a few years with the Alsatian railways and then went to Russia just before the First World War. He was taken prisoner in Kharkov (Charkow) and interned in the Urals when the war broke out. After the war, he taught at Darmstadt, Dresden and Delft; while at the Technical University in Delft (Holland), he did the work on slip-line theory, plasticity and basic rheology, for which he is best known. In 1930, he went to the Massachusetts Institute of Technology as Associate Professor for one year, and in 1931 he taught what must have been the first regular course entitled "Rheology". After that, he returned to Delft and then to Germany and seems to have published less. Interestingly, during the Second World War, he travelled from Germany to an ASME meeting in Philadelphia in 1941 and gave a paper on plate and shell theory, which was printed in the Journal of Applied Mechanics in 1942. He was described only as "Mechanical Engineer, Mainz", and was probably engaged in the war effort. After the second World-War, he worked in German industry. He died in a mountain-sports accident on 6 July 1951.

proposed some influential ideas on coordinates convected with the material (Hencky 1925). These ideas were noted by Oldroyd (1950) and further developed by Lodge (1964). Hencky also invented the logarithmic measure which is now known as 'the Hencky strain' (1929a) and, in a paper devoted to the foundations of hydrodynamics, he also considered rates of change of stress (Hencky 1929b). He produced a Zaremba-like constitutive equation:

$$\frac{\mathcal{D}T_{mn}}{\mathcal{D}t} = \phi_{mn} - \frac{T_{mn}}{\lambda}, \tag{2.20}$$

where λ is a relaxation time, ϕ_{mn} is a function of the velocity gradients and stresses and $\mathcal{D}/\mathcal{D}t$ is the corotational derivative. This was another step in the path towards invariant constitutive relations, but no problems were solved by Hencky; he does not refer to Zaremba's papers.

From now on, we will mainly be describing incompressible materials, and to shorten presentations we will consistently use the notation $\boldsymbol{\sigma}$ for the total stress and $\boldsymbol{T}$ $(\equiv \boldsymbol{\sigma} + p\boldsymbol{I})$ for the extra stress; the pressure is not necessarily (and is not usually) equal to $-\frac{1}{3}tr\boldsymbol{\sigma}$ in nonlinear elastic fluids, but a number of workers in this era made that assumption (e.g. Hencky 1929b; Weissenberg 1931). Since often only rectilinear shearing flows were studied, the consequences of this assumption about pressure were not noticed. For example, Fromm (1947) assumed a bulk elastic law to find the mean pressure, but in simple shearing this does not determine p.

In 1931, Weissenberg noted some of Hencky's remarks about the difference between shearing and irrotational kinematics. Later in 1934, he considered a relation between a deviatoric stress $\boldsymbol{T}$ and a rate-of-strain tensor $\dot{\boldsymbol{S}}$, which was written in the form:

$$\boldsymbol{T} = \boldsymbol{F}(\dot{\boldsymbol{S}}) = \eta_1 \dot{\boldsymbol{S}} + \eta_3 \dot{\boldsymbol{S}}^3. \tag{2.21}$$

Here, $\boldsymbol{T}$ is defined as $\boldsymbol{\sigma} - \frac{1}{3}(tr\boldsymbol{\sigma})\boldsymbol{I}$ and $\dot{\boldsymbol{S}}$ is, as shown by Lodge (1973), after all equal to the usual rate-of-strain tensor $\boldsymbol{d}$. It has to be admitted that Weissenberg's presentation was not as clear as it might have been. He noted that $\boldsymbol{T}$ and $\dot{\boldsymbol{S}}$ must have the same principal directions and it is now clear that the omission of the term in $\dot{\boldsymbol{S}}^2$ is not serious, since, using the Cayley-Hamilton theorem, the term in $\dot{\boldsymbol{S}}^3$ can be replaced by one in $\dot{\boldsymbol{S}}^2$, and similarly for higher powers. As will become clear later, especially following the work of Rivlin (1948a), equations like (2.21) are not adequate to explain observed stress patterns in non-Newtonian liquids.

In this period, one also saw a full three-dimensional tensor description by Hohenemser and Prager (1932) of the inelastic, variable viscosity concept depending on the (three) invariants of the rate of deformation.

In summary, we see, by 1935, many of the essential ideas of a rate-dependent fluid, and the beginning of the systematic use of tensor notation in rheology by Hencky, Weissenberg, Prager and others. This predates Reiner's (1945b) paper, which is sometimes credited with the introduction of tensorial methods in rheology (see, for example, Truesdell, 1952, p225). Finally, the use of the deviator $\boldsymbol{\sigma} - \frac{1}{3}(tr\boldsymbol{\sigma})\boldsymbol{I}$ for the stress $\boldsymbol{T}$ is clearly not satisfactory in general flows, since a further isotropic pressure will still be needed for momentum balance for an incompressible fluid.

In 1933, Hans Fromm in Berlin took up the problem again and in an engineering journal (Ingenieur-Archiv) he set about considering and applying the corotational ideas of Hencky.

He used tensor methods and looked at a number of fairly complex constitutive equations. Most importantly, he considered what we would now call the Corotational Maxwell model:

$$\frac{\mathcal{D}\boldsymbol{T}}{Dt} + \frac{\boldsymbol{T}}{\lambda} = 2G\boldsymbol{d}. \tag{2.22}$$

He proceeded to obtain the correct results for a steady simple shear flow. Specifically, for a Cartesian velocity distribution $v_x = \dot{\gamma}y,\ v_y = v_z = 0$, where $\dot{\gamma}$ is a constant velocity gradient, he obtained the viscometric functions (see Appendix 1):

$$\begin{aligned} \sigma &= \frac{\eta\dot{\gamma}}{(1+\lambda^2\dot{\gamma}^2)}, \\ N_1 &= \frac{G(\lambda\dot{\gamma})^2}{(1+\lambda^2\dot{\gamma}^2)}, \\ N_2 &= -\frac{N_1}{2}, \end{aligned} \tag{2.23}$$

where $\eta = G\lambda$.

Fromm also tried to solve the capillary flow problem for these viscoelastic constitutive equations and, in later publications (1947, 1948), he attempted to interpret theoretically some of Philippoff's experimental results, this being the first recorded attempt to compare nonlinear viscoelastic theory and experiment for a polymeric liquid.

Fromm's (1947) paper also considered pure (irrotational) shearing, which yielded a constant viscosity for the model. The idea of instability in regions of declining stress ($\dot{\gamma}\lambda > 1$) in simple shear implied by (2.23) was also noted. Many of these ideas would be rediscovered more or less painfully in the next forty or fifty years. One must also admire Fromm's persistence in persevering with his pioneering ideas through fifteen years, which included the 1939-45 war.

For completeness, we note that Eisenschitz (1933) and Zaremba (1937) also solved some simple flow problems, but the work attracted little interest at the time.

While it is now conceded that simple corotational derivative models are not very suitable for polymeric liquids, the apparent indifference of later workers to Fromm's work is surprising. Truesdell (1952) in his survey of elasticity and fluid dynamics does footnote the work and gives references, but Noll (1955) only refers to Zaremba explicitly and to Truesdell (1955). Oldroyd (1950) refers to Hencky's (1925) paper on convected coordinates, but not to more recent papers. Weissenberg (1931) did note the Hencky distinction between irrotational and shear flows, but did not refer to the corotational derivative papers of Fromm in any of his papers.

In conclusion, we see that, between 1903 to 1950, the idea of complex stress derivatives occurred to a number of distinguished scientists, but there was a general reluctance to cross-reference earlier work by other authors.

The foregoing discussion has been essentially restricted to *differential* constitutive models, reflecting the subject matter in the papers of relevance. However, it would be remiss of us if we did not draw attention to the excellent paper of Herbert Leaderman (1943), who studied the response of textile fibres and in the process used what appears to be the first nonlinear generalization of Boltzmann's superposition principle, thus providing an early step into nonlinearity and being influential in K-BKZ theory developments (see Chapter 4).

2.5. Karl Weissenberg

Karl Weissenberg was born in Vienna on June 11th, 1893. He studied at the Universities of Vienna, Berlin and Jena, majoring in mathematics, with physics and chemistry as subsidiary subjects. In 1916 he obtained a PhD in Mathematics at Jena, before becoming a Privat-Dozent and later Professor of Physics in the University of Berlin.

Over the years, Weissenberg worked with success in a wide range of disciplines, including mathematics, medical X-rays, X-ray crystallography and, of course, rheology. In 1922, he joined the team of M Polyani at the Kaiser-Wilhelm Institut für Faserstoffchemie in Berlin-Dahlem and over a six year period worked with distinction on X-ray crystallography. He built up an impressive reputation and his efforts, both theoretical and experimental, resulted in the design of an instrument which became known as the 'Weissenberg X-Ray Goniometer'.

In 1933, Weissenberg became a refugee and took up residence in the UK, where he concentrated on his rheological interests. The present historical text gives ample proof of his achievements in the field. So, one reads of the 'Weissenberg (rod-climbing) effect', of the 'Weissenberg Rheogoniometer' for measuring shear and normal stresses, of the 'Weissenberg hypothesis' that $N_2 = 0$ in a steady simple shear flow, and of the 'Weissenberg number' - a dimensionless number to estimate non-Newtonian effects in simple shear flows.

In addition to his own work, Weissenberg was also a motivator of research amongst colleagues and friends, as is clear from the discussion concerning his association with J E Roberts and others during the Second World War. Professor A S Lodge also acknowledges his indebtedness to Weissenberg for stimulating his interest in normal-stress measurement in the early 50s.

Karl Weissenberg had the reputation of 'being an entirely engaging and unselfish person', of 'being a delightful companion, an ever helpful friend and also an excellent tennis player'. An obituary, written at the time of his death in 1976, concluded that 'he was notable in his scientific achievements and noble in his personal qualities'.

Chapter 3

Interlude: Rheology Becomes an Independent Science: Societies, Congresses and Journals

3.1. Introduction

1929 was a vintage in the history of rheology. It saw the formal introduction of the term rheology, witnessed the founding of the first national Society of Rheology (in the USA) and the creation of the first scientific journal devoted exclusively to rheology (the Journal of Rheology).

That is not to say that no rheological research was done before 1929. Indeed, Professor E C Bingham of Lafayette College, Easton, Pa., whose vision had much to do with the events of 1929, had carried out research on the plasticity of concentrated suspensions well over a decade earlier, to say nothing of the contributions of famous scientists like Maxwell, Kelvin and Boltzmann in the 19th Century. The words of Scott Blair (1972) are apt: "Just as Molière's famous Monsieur Jourdain found that he had been writing prose all his life without appreciating what he was doing, so a number of chemists, physicists and others practised rheology before the name was invented or rheological organization established".

But the time did come to recognize the distinctive character of the evolving field. So, at a Plastics Symposium in the US, a committee was formed to investigate the formation of a permanent organization to look after "Rheology". The committee met on April 29th, 1929, and the first meeting of the Society took place on December 19th and 20th, 1929.

It was agreed that the organization be known as the Society of Rheology and its official publication as the Journal of Rheology, which was to be published quarterly. It was further agreed to hold meetings annually.

The formal definition of rheology is invariably attributed to **Bingham**, who was apparently assisted by a colleague who was a classics scholar. The first issue of the Journal of Rheology contains the relevant background and definition:

> *THE NAME*
>
> *The term deformation and flow of matter is a rather cumbersome one to cover the subjects of elasticity, viscosity and plasticity. There is no single word to*

cover the field, so the only recourse has been to invent one. The Greek roots to flow (ῥέω) and science (λόγος), already familiar in numerous words such as rheostat and geology, made the term rheology appear to be at the same time distinctive and self explanatory.

By common consent, rheology excludes such subjects as pure hydrodynamics and the classical theory of elasticity.

The first meeting of the Society of Rheology took place at the National Bureau of Standards in Washington. Interestingly, it was attended by two of the three scientists who, each in their own way, were to play a significant role in the evolving science. Bingham was one of them and **G W Scott Blair** of the UK was the other. Unfortunately, the third member of the influential triumvirate, Professor **Marcus Reiner**, who had figured prominently in early discussions with Bingham, had to be in Palestine at the time, although he did publish two papers in the first issue of the Journal of Rheology (Reiner 1929a,b).

The early issues of the Journal of Rheology contained papers of quality and three volumes were published, until in Vol. 3 No 4 there was a note to the effect that The American Institute of Physics had taken over the editorial mechanics and business management of the Journal of Rheology. However, this development was overtaken by events, influenced presumably by the Great Depression, and the December 1932 issue was the last.

A Journal of the Society of Rheology was restarted in 1957 with the name 'Transactions of the Society of Rheology', However, the title reverted to the old title 'Journal of Rheology' in 1977/78.

3.2. Developments in Europe

Following the successful launch of the Society of Rheology in Washington in 1929, there was a long period of inactivity elsewhere. Bingham and Reiner had hoped that the American Society would become international (see, for example, Scott Blair 1982), but this did not happen and it was not until the pressures of the Second World War that further groups were formed, notably in the U.K. The British Rheologists Club was founded in 1940 through the efforts of a triumvirate consisting of G W Scott Blair, who was working at the National Institute for Research in Dairying at Reading, V G W Harrison and H R Lang, the Secretary of the Institute of Physics. Interestingly, they were brought together by the similarity between the rheology of printing inks and those of the secretion from the uterine cervix of the cow! Lang arranged for Harrison and Scott Blair, who were independently working on methods of measuring the properties of these materials, to meet and discuss common aspects of their work and this can be viewed as the first meeting of the British Rheologists Club (see for example, Wookey 1961). Lang (1946) graphically describes the early meetings:

"It was a fine summer afternoon in 1940 and, feeling somewhat exhausted by the difficult war conditions on evacuation to Reading, I went a little way into the country with my wife. Begrudging the few hours away from the temporary office at the Institute of Physics, I consoled myself by fulfilling a long-standing promise to visit Scott Blair's laboratory at the National Institute for Research in Dairying, at Shinfield, some three miles out of

Reading. He explained to me the difficulty he was having in devising suitable instruments for measuring the rheological properties of the secretions from the uterine cervix of the cow, which he thought might prove a useful early test for pregnancy - a matter of great importance to the dairy industry during the war. It immediately occurred to me, from what he said, that this problem was very similar to that of V G W Harrison, who was trying to measure the rheological properties of printing inks. In the normal course of my work as Secretary of the Institute of Physics, I therefore gladly arranged for these two to meet. That meeting must be taken as the first meeting of the Club....".

"We three appointed ourselves a Committee having decided to form a Club, which should have no rules and no formalities - indeed a Club, where one should speak as freely as in the lab. The committee was to meet either in my attic room in the University of Reading or in Scott Blair's drawing room".

A circular letter dated 1940 was sent to scientists interested in problems of deformation and flow and an inaugural meeting on "Rheology in Industry" was held on November 16th, 1940, at the University of Reading, the first president being Sir Geoffrey Taylor FRS, a leading British applied mathematician.

The first major conference organised by the British Rheologists Club was held at St Hilda's College, Oxford, in July, 1944.

Volume 1 of the Bulletin of the British Rheologists Club appeared in January 1941. The Bulletin contained a set of "Abstracts" of relevant papers, which were a useful source of information for the members of the club, of which there were 80 or so at the time. Later, abstracts were given their own publication and "Rheology Abstracts" came into being in 1958.

The British Rheologists' Club became "shockingly successful" and by 1945 membership had reached 300. In 1950, after much heart searching, it was decided to change the name of the club to the British Society of Rheology, the main reason for this seems to have been the implications attached to the word 'Club' by some senior industrialists, who raised suspicious eyebrows when their staff wished to take a few days off to attend 'Club' meetings!

In early 1947, the Secretary of the British Rheologists' Club approached Dr R.Houwink, a Dutch rheologist, concerning the desirability of holding an International Congress on Rheology in the Netherlands. The suggestion was received very favourably by Dutch rheologists and Professor J M Burgers and Dr Houwink made a provisional plan, which they brought before the Committee for the Study of Viscosity and Plasticity of the Royal Netherlands Academy of Sciences at Amsterdam. This committee agreed to sponsor the Congress and, in September 1947, the newly-formed "Joint Committee on Rheology" of the International Council of Scientific Unions did likewise. Burgers was appointed Chairman of the Organizing Committee and Houwink the First Secretary. Other notable committee members were H Kramers and A J Staverman.

Scheveningen was chosen as the most suitable venue for the Congress and September 21st-24th, 1948 as the most appropriate timing. As a deliberate policy decision, it was decided that a substantial part of the meeting should be devoted to General Lectures and amongst those invited were F R Eirich, M Reiner, K Weissenberg, G W Scott Blair and A L Copley.

The Congress was attended by about 250 participants (130 from Holland and 120 from other countries, amongst whom 80 were from Great Britain). The character of the Congress was purposely kept informal; apart from a general dinner, little attention was given to social entertaining, although the Congress was followed by an excursion.

An exhibition of rheological apparatus was organized during the Congress, a successful feature that was to become standard in the Congresses that followed. There was also a discussion on rheological nomenclature. The Proceedings of the Congress (1949) provides an interesting account of the background administrative details leading up to the meeting as well as an excellent overview of the state of rheology, especially in Europe, in the late 40s.

The success of the first Congress ensured that a second would be held and this was duly organized by the British Society of Rheology in 1953. The venue was Oxford and depending on the counting procedure, estimates of the number of attendees range from 240 to 400. Scott Blair was the Secretary and he carried the major responsibility for the meeting's organization; Harrison (1954) edited the Proceedings. Sir Geoffrey Taylor was the President and opening speaker with an address entitled *"Rheology for mathematicians"*. The papers were read, and discussions took place, in English, French and German. Marvin (1976) in his "Brief History of the International Committee on Rheology" highlights the fact that nearly half the contributed papers involved *polymeric* materials, beginning a trend which is still in evidence today. Indeed, it is clear that there is a strong link between the emergence of rheology as an important science and the flourishing of the polymer industries (see, for example, Markovitz 1985).

It was at the Oxford Congress that the International Committee on Rheology was formed, with the following remits: (1) To be a permanent body for organizing future International Congresses on Rheology, (2) to encourage new national organizations for the study of rheology, and (3) to act as coordinating body in other international cooperations in the field of rheology (Scott Blair 1953, Marvin 1976).

Continuing the initial procedure of holding Congresses every five years, the third was organized at Bad Oeynhausen, West Germany in 1958.

The German Society of Rheology (Deutsche Rheologische Gesellschaft, DRG) had been founded in (West) Berlin in October 1951. However, a little later, a second rheological organization "The German Rheological Association" (Deutsche Rheologen Vereinigung, DRV) was founded. This had a somewhat different flavour with an industrial bias; (W Meskat of the Bayer Company, Leverkusen, was prominent in this development).

The two organizations existed side by side for many years, but this dichotomy was driven as much by immiscible personalities as fundamental scientific concerns. However, in 1975, the two Societies merged, maintaining the name "Deutsche Rheologische Gesellshaft (DRG)".

Professor Kurt Kirschke, who for many years worked at the Bundesanstalt fuer Materialpruefung (BAM), Berlin, played a prominent administrative role in German rheology spanning three decades and, during that time, he was responsible for the abstracting service "Documentation Rheology", until its demise in 1987 due to lack of funding.

3.3. On Rheological Journals

The 1958 German Congress provided a stimulus for the formation of an international journal: Rheologica Acta. However, it was far from being the only factor of importance, since the basic need for rheology to have its own journal was shared by a growing number of German scientists. For economic reasons, the journal was at first coupled with the well established 'Kolloid-Zeitschrift', later called the 'Journal of Colloid and Polymer Science'. So, in 1958, Rheologica Acta began as a 'Supplementary Journal' to Kolloid-Zeitschrift. The first issue contained original articles, but the second and third issues, published in August 1958, contained preprints from the 3rd International Congress. There followed a publication delay caused by financial problems, before the remainder of the Congress proceedings appeared in March 1961.

Although Dr W Meskat, the then secretary of the DRV, was the most prominent player in these journal developments, there was a significant input from others, notably Professor Kroepelin, the President of DRV, and also Professor F H Müller, the International Congress President.

At the 1963 International Congress at Brown University, representations were made to establish Rheologica Acta as *the* international journal of rheology, but this was opposed by American rheologists in particular, who saw the development as a threat to the future of the Transactions, and the initiative came to nought. Nevertheless, under Meskat's Editorship, Rheologica Acta soon established itself as an important international journal. A second crisis arose in the 73/74 period, due to an overambitious promise to publish the proceedings of the Lyon International Congress. This so overburdened the Journal that many prospective authors became disgruntled at the long publication delay. The crisis was averted when the new Executive Editor, Professor H Giesekus, reached agreement with the publishers in 1975 to bring out Volume 14 with 12 issues.

Giesekus was an influential Editor of Rheologica Acta until his retirement in 1989. He was succeeded by Professor H H Winter[1] of the University of Massachusetts at Amherst.

For many years, Rheologica Acta and the Journal of Rheology provided the only dedicated channels for the dissemination of rheological research, although, not surprisingly, many authors patronized journals serving mainline disciplines like Mechanics, Engineering and Processing. A *third* dedicated journal was to come into being in 1976. Elsevier, the publishers, had been advised of a need in the area of non-Newtonian Fluid Mechanics and the first issue of the Journal of non-Newtonian Fluid Mechanics (JNNFM) appeared in 1976, under the Editorship of Professor Ken Walters of Aberystwyth. The advent of the new journal did not generate unbridled enthusiasm, the main criticism coming from Professor George Batchelor, the Executive Editor of the influential Journal of Fluid Mechanics (JFM). Batchelor, rightly or wrongly, felt that non-Newtonian fluid mechanics ought to be viewed as a branch of fluid mechanics and that, by implication, papers in the area should be submitted to JFM or at least a supplementary journal of JFM.

[1]Horst Henning Winter studied at the University of Stuttgart, Germany, where he obtained a PhD degree in Chemical Engineering in 1973. Earlier, in 1968, he had received an M.S. in Chemical Engineering from Stanford University. Winter moved to the University of Massachusetts at Amherst in 1981 to begin a successful academic career, and in 1994 he was formally awarded the title of "Distinguished University Professor" by the University's Board of Trustees. He received the S.O.R. Bingham medal in 1996. He is best known for his research on the characterization of the rheology of polymers during gelation.

So, for example, in a review of new journals, Batchelor (1975) wrote that, with regard to the JNNFM, "the dangers of separateness and fragmentation come to mind". This point was taken up by one of the members of the original JNNFM Editorial Board, Professor Gianni Astarita,[2] and in an article entitled "Is non-Newtonian Fluid Mechanics a culturally autonomous subject?" we read (Astarita, 1976) "it is not a statement to be taken lightly, either on its own grounds or in view of by whom and where it was voiced; a critical self-examination is in order".

Astarita reviewed the arguments and finally concluded "Of course, one may regard 'fluid mechanics' as encompassing both NFM and NNFM - this is a matter of personal choice; but, if one does take such an attitude, then NFM is such a special subcase as to be degenerate... In conclusion, the critical examination which I have tried to carry out leads me to the belief that the *Journal of non-Newtonian Fluid Mechanics* does not induce fragmentation in the landscape of the technical literature, but on the contrary can and should be a focus of aggregation of a culturally homogenous literature, which so far has been dispersed in a large number of different journals.

The passage of time has added weight to Astarita's arguments and the Vol 50 milestone of JNNFM was passed in 1993.

So, since 1976, rheology has had *three* dedicated journals. These have sought to serve the field in a complementary, rather than competitive, fashion and, as a tangible expression of this, the Executive Editors of the Journals have in recent years served on all three Editorial Boards.

3.4. More Congresses

The fourth and fifth International Congresses took place in Providence, Rhode Island, USA and Kyoto, Japan, in 1963 and 1968, respectively. By 1968, activity in the field had grown to such an extent that it was decided to make the interval between successive International Congresses four years instead of five and Congresses followed in Lyon, France (1972), Gothenburg, Sweden (1976), Naples, Italy (1980), Acapulco, Mexico (1984), Sydney, Australia (1988), Brussels, Belgium (1992) and Quebec, Canada (1996); the 2000 Congress is scheduled for Cambridge, England.

The 7th Congress was originally planned for Prague in 1976. However, 'local' difficulties meant that the Czech organizers were unable to undertake the sponsorship of the Congress. A fall-back situation agreed on at the 6th Congress in Lyon was activated and Gothenburg, Sweden, became the alternative venue for the 7th Congress.

For completeness and (hopefully) interest, we provide the background information for the International Congresses, together with one or two subjective memories drawn from reports of the various Congresses.

---

[2]Giovanni (Gianni) Astarita earned his Masters degree in Chemical Engineering from the University of Delaware and his doctorate from the University of Naples in 1957. He then joined the engineering faculty at Naples to begin an illustrious academic career, which was to include a significant rheological component, but was by no means restricted to that field. He also had a long association with the University of Delaware and he was a visiting professor there annually in the fall semesters from 1973 to 1995.

Gianni Astarita received many honours during his illustrious career. One that gave him much pride was his election in 1994 as a Foreign Associate of the National Academy of Engineering of the United States. He died suddenly on April 24th, 1997, at the age of 63.

4th Congress August 26-30, 1963

Location: Brown University, Providence, Rhode Island.
Co-chairmen: R S Marvin and R S Rivlin.
Comments: During the Congress, the Bingham medal of the (American) Society of Rheology was awarded to Professor C Truesdell. "There was a memorable gastronomical climax to the meeting - a clambake at Mystic Seaport".
Dr M Mooney attended this Congress, together with many who would subsequently star in the field. The Proceedings were printed in four volumes and appeared about a year after the Congress.

5th Congress October 7-11, 1968

Location: Kyoto, Japan.
Background: The first Reoroji Töronkai (Rheology Symposium) was held at the University of Tokyo in November 1951. Various administrative developments took place in the ensuing years before the Society of Rheology, Japan, was officially established on the 1st January 1973.
Earlier, in 1968, the International Congress was hosted by *two* organizations with rheological interests, associated with the Society of Polymer Science and the Japanese Society for Testing Materials. The 1973 launch of the independent Japanese Society of Rheology removed the anomaly (Onogi 1983).
Co-chairmen: M Horio and B Tamamushi.
Attendees: Total 620 (including 73 accompanying persons) from 21 countries. 403 (including 17 accompanying persons) were Japanese nationals.
Comments: "About half the papers were devoted to polymers and polymer solutions". "On the first evening of the Congress, the President of the Japanese Scientific Council held a reception in the Kyoto Tower Hotel. There the Western guests valiantly attempted to drink sake from little wooden boxes and to eat a range of delicacies using chopsticks, but our hosts had thoughtfully provided alternative utensils for those who needed them. Geishas were present to look decorative and to keep the beer glasses filled, and a programme of dances was given by a group of maiko.."

6th Congress September 4-8, 1972

Location: Université Claude-Bernard, Lyon, France.
Organizers: G Vallet, B Persoz and M Joly; Secretary C Smadja. A Larcan was the secretary of the Biorheology section.
Attendees: Total 600. French 190, Americans 100, British 65.
Comments: 275 papers were presented on General Rheology, 80 on Biorheology.

The Proceedings for this Congress eventually appeared in Rheologica Acta.
"Another drawback at the University Residence was the absence of hot water!"
"It is hoped that some good resulted from queuing each day for lunch and listening for an hour to votes of thanks in three languages *before* the closing banquet".

7th Congress August 23-27, 1976

Location: Chalmers University of Technology, Gothenburg, Sweden.
Organizers: J Kubat, E Forslind, L-E Gelin, F K G Odqvist; Secretary C Klason.
Attendees: 500 from more than 20 countries.
Comments: The relatively new concept of 'poster sessions' was tried out at the Congress, a third of the papers being scheduled for presentation in this way. To the credit of the organizers, the Proceedings were available at registration for the first time, a practice adopted by all succeeding Congresses.
"It is rumoured that the organizers had 5000 copies of the Proceedings printed, of which 4000 remain unsold!"

8th Congress September 1-5, 1980

Location: University of Naples, Italy.
Organizers: G Astarita, G Marrucci and L Nicolais.
Attendees: 519, plus 226 accompanying persons, from 35 countries.
Comments: "An early breakfast, a ride to the harbour in Sorrento and a sea crossing to Naples; this was the start of a day at the eighth International Congress. And what a masterpiece of organisation it was - highlighted on the first day by an escort of police outriders through the streets of Naples!" "The conference dinner, which took the form of a buffet, resembled an invasion by a plague of locusts; rheologists clearly have insatiable appetites."

9th Congress October 8-13, 1984

Location: Acapulco Mexico.
Organizers: B Mena, A Garcia-Rejón and C Rangel-Nafaile.
Attendees: 450.
Comments: "The scene at the opening plenary lecture by Professor Ronald Rivlin may be captured by imagining a small figure on a large stage in front of an extremely large screen marshalling rows of symbols with authority. He has a pointer which does not reach half-way up a display of tensorial equations; the microphone catches the remark "Most people wouldn't touch this stuff with a ten foot pole" and the audience agrees".
The organizers recall the occasion when Professor Joachim Meissner of ETH, Zürich, insisted on having two simultaneous slide projectors, a film projector

and a video recorder all at the same time at 8 o'clock on the morning of his talk!

"Less conventional was the conclusion of the final banquet, with the newly elected president of the International Committee on Rheology (Balta Mena) leading his rock-and-roll band with enthusiasm".

The annular swimming pool, with the central cashless bar, was a notable feature of the informal part of this Congress.

10th Congress August 14-19, 1988

Location: University of Sydney, Australia.

Background: In late '58, Dr Paul Grossman approached the Secretary of the British Society of Rheology about rheological activities in Australia. He was informed that there were at that time two or three members of the British Society of Rheology resident in Australia and that the formation of an Australian Branch of the BSR would be welcomed.

On his return to Australia, Grossman contacted a number of interested parties and had a positive response from, amongst others, Leopold Dintenfass, then working in the paint industry before switching to biorheology, and Nick Tschoegl, then at the Bread Research Institute and later to become Professor at Cal Tech in the US and, for many years, Secretary of the International Committee on Rheology. Dintenfass organized a meeting in October 1959 and a NSW Branch of the BSR was established.

Driven by support from CSIRO, a rheology meeting was held in Melbourne in February 1960 and this resulted in the formation of a Victorian Branch of the BSR.

By the end of the 60s, interest in Rheology had flagged in NSW and the Branch was unable to maintain itself. As a result, an Australian Branch of the BSR was formed, with its committee mainly located in Victoria.

After that time, Australian rheology flourished and, at the 3rd National Conference in Melbourne in 1983, Professor Roger Tanner suggested that the time was ripe to form an independent Australian Society of Rheology and this came into being in late 1983, with the full support of the BSR.

Chairman: R I Tanner, Secretary: N Phan Thien.

Attendees: 350.

Comments: "Most people stayed at St John's and Sancta Sophia Colleges or in the Camperdown Travelodge across the road. The colleges were quite cold in the evenings - buildings of this type do not usually have central heating - but modern fan heaters were standard issue to the residents". However despite the fact that the Congress was held in the winter season, August, the temperature reached 24°C most days, with much colder nights. The night of the Harbour Cruise dinner was warm and calm. (This was fortunate, since the following week was cold, wet and windy!)

11th Congress August 17-21, 1992

Location: Palais des Congres.
Co-chairmen: M J Crochet and J Mewis, Secretary: Paula Moldenaers.
Attendees: 600 from 38 countries.
Comments: "The printed proceedings of 1020 + xliv pages, edited by P Moldenaers and R Keunings, and beautifully bound in hard covers by Elsevier, weighed a total of 3kgs!"

12th Congress August 18-23, 1996

Location: Québec, Canada.
Chairman: D DeKee, Secretary: M Dumoulin.
Attendees: 586 plus 85 accompanying persons.

3.5. The International Dimension

As the science of rheology has grown in its constituency and influence, there has emerged, quite naturally, three block groupings, one centred on the Society of Rheology in the US and taking in the Americas; another associated with Europe; and finally, and more recently, a Pan Pacific grouping, which reflects in some measure the growing economic importance of such countries as Japan, Korea and China.

The meetings of the Society of Rheology which take place annually at various North American locations have long been important events in the rheological calendar and they attract large attendances both from within and outside the US. These meetings reflect the continued prominence and influence of American rheological activity.

Postwar political developments in Europe, culminating in the formation of the European Union and other wider economic groupings, have been mirrored by attempts to forge closer links between the national Societies of Rheology with a view to creating a cohesive European activity in the field. One manifestation of this has been the organization of European Rheology Conferences, which take place every four years, suitably spaced between the International Congresses. The early initiative for this development came from Professor K Kirschke of Berlin and Professor J Schurz of Graz, Austria, and the first conference was duly held in Graz in 1982. Since that time, further conferences have been held in Prague (1986), Edinburgh (1990) and Seville (1994).

A further reflection of the European element has been the formation in 1996 of the European Society of Rheology, an umbrella organization, with Professor K Walters (UK) the first President, Professor J M Piau (France) the Vice President, Professor M H Wagner (Germany) the Secretary and Professor J Mewis (Belgium) the Treasurer. One of the first actions of the fledgling Society has been to institute a 'Weissenberg medal', which will be the European counterpart of the Society of Rheology Bingham medal.

More recently, the growing Pacific-rim influence in the field has resulted in the organization of Pan Pacific Conferences, the first of which was held in Japan in 1994. The second was held in Melbourne, Australia in 1997.

3.6. Eugene C Bingham

Eugene Cook Bingham was born in Cornwall, Vermont, on December 8th 1878. He graduated from Middlebury College in his home state in 1899, before moving to the Johns Hopkins University, where he obtained a PhD degree in 1905. He then spent a year studying at the Universities of Leipzig, Berlin and Cambridge.

From 1906 to 1915, Bingham held the position of Professor of Chemistry at Richmond College, Vermont, and then from 1915 to 1916, of Assistant Physicist at the US Bureau of Standards, where he worked on problems in viscous flow.

In 1916, Bingham moved to Lafayette College in Easton, Pennsylvania, where he spent the remainder of his academic life. He retired from the Department Headship in 1939 to become Research Professor, a post he held until his death on November 6th, 1945.

Between 1906 and 1914, Bingham wrote a series of papers on 'viscosity, and fluidity' and then a paper entitled *"Plastic flow"* appeared in 1916 (Bingham 1916). There followed a series of papers on viscosity and plasticity and on viscometers and plastometers, and in 1922 the famous book *"Fluidity and plasticity"*.

As the general text makes clear, E C Bingham was largely responsible for the formation of the (American) Society of Rheology and 'rheology' was to figure prominently in his research publications from 1929 onwards.

Obituaries at the time of his death (see, for example, The Octagon, Jan 1946, Vol 29 No 1) refer to Bingham's other passions, such as blazing a trail along the Blue Mountains and of seeking a solution to the problem of highway illumination.

Bingham was clearly a man of positive views, great energy and enthusiasm. He was never known to lose his temper or to show impatience at the weakness of others and he was uniformly kind in his judgements.

3.7. G W Scott Blair

George William Scott Blair was born in Weybridge, England in 1902. He read Chemistry at Trinity College, Oxford, where his tutor was Sir Cyril Hinshelwood. There followed two years industrial experience as a Colloid Chemist at Henry Simon Ltd, Manchester, where he worked on the viscosity of flour suspensions.

In 1927, Scott Blair moved to the Physics Department of the Rothamsted Experimental Station, where he remained for ten years. The main emphasis of his work there appears to have been in soil science, but he was still able to devote some time to the rheology of doughs and other materials.

Scott Blair moved to the National Institute for Research in Dairying, University of Reading, in 1937 and remained there until his retirement in 1967. Initially, he was Head of Chemistry at the NIRD, but later he moved to the Physics Department. During thirty very productive years, Scott Blair worked on the rheology of dairy products; he also studied the properties of bovine and human cervical mucus and many other materials. Following his retirement, he worked part-time on blood coagulation at the Oxford Haemophilia Centre and also consulted on rheological matters, especially the rheology of foodstuffs.

In all, Scott Blair published six books and over 250 research papers in a career which spanned over 45 years. He was a founding Co-Editor of the Journal of Biorheology, along with his close friend, A L Copley.

Scott Blair's research papers covered many areas of conventional rheology, as well as biorheology and psychorheology. His interests were therefore very broad, although it is clear that he was not particularly at home with some aspects of theoretical rheology. For example, on one occasion in 1972, he wrote "There is, as yet, no general theory, but many branches of higher mathematics are now being applied to rheology.... Unfortunately, there is also some danger of misapplication of mathematics and the introduction of unnecessarily complex mathematical methods".

The general text in the present book provides ample evidence of Scott Blair's massive contributions to the formalization of rheological activities both in the US and the UK, and we recall that he was Secretary of the 2nd International Congress on Rheology held at Oxford in 1953.

Scott Blair was President of the British Society of Rheology from 1949-1951 and received a Founders Gold Medal from that Society. He was also awarded the Poiseuille medal of the International Society of Haemorheology (now "Biorheology") and the Freundlich Memorial Medal of the Deutsche Rheologische Gesellschaft (DRG).

In addition to his passion for rheology, Scott Blair also had interests in languages, philosophy and music. For many years he was a keen choral singer in oratorios, along with his devoted wife, Rita.

George Scott Blair was the archetypal British gentleman. He was always friendly and charming and never seemed to lose his temper; there is no evidence of him ever speaking ill of anyone. His latter years were dogged by ill health, but even with failing eyesight and diminishing physical powers, he retained his outward composure. He died in 1987.

The extensive 'Scott Blair Collection of papers on Rheology' is housed in the Hugh Owen Library of the University of Wales, Aberystwyth.

3.8. Marcus Reiner

Marcus Reiner was born on January 5th, 1886, of Jewish parents in Czernowitz, which was then a part of the Austro-Hungarian Empire. He studied Civil Engineering at the Technische Hochschule in Vienna and, after obtaining his degree, he moved first to Berlin and then to Essen. Later, he returned to Czernowitz and, during the First World War, he was engaged in such pursuits as the reconstruction of damaged bridges in the Balkans.

After the war, Czernowitz became a part of Romania. By this time, Reiner had become a convinced Zionist and he decided to emigrate to Palestine, which was then a British Mandate.

As was normal at the time, Reiner's stay in Palestine began with hard manual labour, but the British soon appreciated his engineering skills and, up until the ending of the Mandate in 1948, he worked for them.

In 1927, Reiner visited Easton, Pennsylvania, and interacted with Professor E C Bingham. They decided that, if a chemist (Bingham) and an engineer (Reiner) had so many problems in common, a special branch of science should be founded. The general text provides the relevant details of the outcome.

Some of his early work was carried out in association with Mrs R Riwlin, whose career was brought to a tragic end by a road accident. Interestingly, R S Rivlin (now spelt with a 'v'), who was to play such a prominent role in post Second World War developments, was a nephew of Mrs Riwlin. Unlike his aunt, R S Rivlin had no formal association with Reiner, but in the late forties they sometimes worked on similar research topics, one of which led to the construction of the so called Reiner-Rivlin fluid (see §4.1).

After the ending of the Mandate in 1948, Reiner became a Professor at the Technion in Haifa, Israel. He wrote a number of text books (including the influential *'Twelve lectures'*, 1949)and many research papers and some of these were bound together in book form not long before his death (Reiner 1975).

To celebrate his 80th birthday, a book was compiled by his colleague, Dr David Abir (1969), to which many notable rheologists contributed. At the same time, the British Society of Rheology presented Reiner with a Gold Medal.

Throughout his life, Reiner was known for a 'delightful sense of humour and fund of good stories'.

Reiner died three months after his 90th birthday on April 25th 1976. Fortunately, he was in good enough health to be present and to appreciate a ceremony held a few weeks before his death. This was attended by the President of the State of Israel and at it a "Marcus Reiner Chair of Rheology" was founded, to be located at the Engineering Department of the Technion. Not long after the Ceremony, Reiner had a stroke; he died shortly afterwards, his life-work completed (Scott Blair 1976).

3.9. Picture Gallery

One of a group of photographs presented by the Deutsche Rheologische Gesellschaft to Dr G W Scott Blair in his capacity as Organizing Secretary of the 2nd International Congress on Rheology in Oxford, 1953. From left to right: Mr McCutcheon and Dr Haid (Berlin), Dr Kubat (Stockholm) and Professor Gross (Rio de Janiero).

Further pictures from the 1953 DRG Scott Blair collection showing (on the left) two prominent European rheologists of the time: Dr M Joly (Paris) and Professor M Pfender (Berlin). On the right, Dr H Umstätter (Berlin) and Professor B Gross (Rio de Janiero) (partly hidden by the blackboard) discuss an issue in 'the theory of linear viscoelasticity'.

1966. Professor Marcus Reiner giving an acceptance speech after receiving a BSR gold medal from the then President of the British Society of Rheology, Professor A J Kennedy. Looking on attentively is Reiner's long time friend, Dr G W Scott Blair.

Photograph taken at the International Congress on Rheology in Bad Oeynhausen, West Germany in 1958. From left to right: J G Oldroyd, A B Metzner, R L Whitmore, B A Toms, Mrs Toms.

Dr Karl Weissenberg illustrating a lecture in Moscow (c1965). Looking on is Professor Alexander Malkin and a Russian interpreter.

A meeting of ICR delegates at the Kyoto Congress, 1968. Professor J D Ferry is in the Chair.

Professor Joachim Meissner addressing a meeting of ICR delegates at the Brussels Congress, 1992.

1969. Professor G V Vinogradov, flanked by Professor D Dowson and Professor G R Higginson (BSR gold medal recipients) and backed by the BSR President, Dr L Grunberg, makes a firm point in an impromptu after-dinner speech.

H Münstedt, M H Wagner, J Meissner, F N Cogswell, J R A Pearson and H M Laun in relaxed mode on the occasion, in 1981, when Münstedt, Wagner, Meissner and Laun jointly received the Annual Award of the British Society of Rheology for their work on Extensional Flow.

Photograph taken in 1977 of F M Leslie and J L Ericksen, well known for their seminal work on the theory of anisotropic fluids.

An early photograph of A E Green and R S Rivlin, taken during their fruitful period of collaboration on constitutive equations for materials with memory.

The participants at a meeting on "Problems in nonlinear continuum mechanics" held at the Dublin Institute for Advanced Studies in May 1980. The group contains many well known theoretical rheologists, including (seated) F M Leslie (1st from left), A J M Spencer (3rd from left), R S Rivlin (5th from left), D D Joseph (6th from left), A C Pipkin (7th from left) and (standing) M A Hayes (second row, 2nd from left). Seated 4th from left is J L Synge, best known for his work on relativity.

The committee of the European Society of Rheology at the inaugural meeting in Leuven, Belgium, March 4th 1997. From left to right, J Mewis (Treasurer), J Ferguson (UK), G Marrucci (Italy), R Gaudu (France), E Mitsoulis (Greece), Y Ivanov (Bulgaria), C Klason (Nordic Society), K Walters (President), M H Wagner (Secretary), I Emri (Slovenia), J M Piau (Vice President), C Gallegos (Spain), P Moldenaers (Belgium), H E H Meijer (Netherlands).

Professor G Astarita (Italy), Professor J Schurz (Austria) and Dr D R Oliver (UK) in relaxed mood during the 2nd European Rheology Conference in Prague (1986).

1987. Photograph taken in Dortmund at an event to mark the retirement of Professor H Giesekus. From left to right: A B Metzner, K Walters, Mrs Hanna Giesekus, H Giesekus and (partly visible) H Janeschitz Kriegl.

1971. Professor A S Lodge being presented with the Bingham medal by Professor H Markovitz at the 42nd Annual Society of Rheology meeting at the University of Tennessee. Mrs Helen Lodge (partly hidden) is seated on the right.

Professor R B Bird receiving the National Medal of Science in the Rose Garden of the White House in 1987 from President Ronald Reagan.

1986. Professor J R A Pearson being presented with a Gold Medal of the British Society of Rheology by the President, Professor I M Ward.

1986. Dr B A Toms being presented with the Annual Award of the British Society of Rheology by the President, Professor I M Ward. (From the BSR Bulletin, 1986, Vol 29, No3.)

1990. Dr V G W Harrison, one of the founders of the British Rheologists Club in 1940 and recipient of a BSR Founders Gold Medal in 1970, speaking at the European Rheology Congress Banquet in Edinburgh, Scotland.

1990. A subset of the top-table group at the Edinburgh European Rheology Congress Banquet. From left to right: Mrs H Giesekus, Professor Sir Sam Edwards, Professor J Ferguson (BSR President), Professor H Giesekus and (partly hidden) Mrs J Ferguson. Professors Edwards and Giesekus were presented with BSR Gold Medals at the Congress.

1980. Professor J Mewis of Leuven in deep conversation with Professor M M Denn of Berkeley on the open deck of one of the ferries which transported participants from the Sorrento hotels to the ICR scientific sessions in Naples.

1992. 11th International Congress on Rheology, Brussels, Belgium. Dr H M Laun of BASF, Germany, talking to one of the Congress organizers, Professor M J Crochet, during a coffee break.

Chapter 4

Constitutive Equations

4.1. Inelastic Fluids

A major event in the development of constitutive equations centred around the publications of Stokes (1845) in the middle of the nineteenth century, in which he proposed a very general concept of viscosity. Specifically, Stokes proposed a constitutive equation of the form:

$$\boldsymbol{T} = \boldsymbol{f}(\boldsymbol{d}). \tag{4.1}$$

This equation anticipated the fact that the rate of strain tensor $\boldsymbol{d}$ was the appropriate kinematic variable, with the attendant implication that the vorticity should not appear explicitly in (4.1) (cf. Truesdell 1960a).

In his work, Stokes restricted attention to the case in which $\boldsymbol{f}$ is a linear function of $\boldsymbol{d}$; it was left to Reiner (1945b) and **Rivlin** (1948a) in the middle of the present century to remove this restriction.

Truesdell (1965a) views the paper by Reiner (1945b) as heralding the rebirth of continuum mechanics as a rational science, although it is admitted that, as so often happens, the results of the paper were better than the assumptions and methods used to describe them.

Reiner (1945b) attempted to provide a phenomenological theory for so called dilatant materials, which would explain the 'wet-sand effect' (wet sand when deformed appears to be drier). Rivlin (1971), echoing some of the sentiments of Truesdell, noted that much of Reiner's paper is incorrect, but he does acknowledge that the use of the Cayley-Hamilton theorem (a matrix satisfies its own characteristic equation) allowed Reiner to express the constitutive equation in the relatively simple form:

$$\boldsymbol{\sigma} = c_0\boldsymbol{I} + c_1\boldsymbol{d} + c_2\boldsymbol{d}^2, \tag{4.2}$$

where c_0, c_1 and c_2 are functions of the three invariants of $\boldsymbol{d}$. At the time, Reiner was apparently unaware of normal-stress effects and accordingly proposed unwarranted restrictions on these functions.

A little later, it appears that Weissenberg visited Rivlin and mentioned normal-stress effects. This aroused Rivlin's curiosity and interest, and he finally found time to follow up the interest during a year's stay at the National Bureau of Standards in Washington.

In Rivlin's (1971) own words "I developed a theory for the flow of viscoelastic fluids based essentially on the assumption that the Cauchy stress is a function of the velocity gradients only and that the fluid is isotropic and incompressible". This led to the constitutive equation:

$$\boldsymbol{\sigma} = -p\boldsymbol{I} + c_1\boldsymbol{d} + c_2\boldsymbol{d}^2, \tag{4.3}$$

where p is an arbitrary hydrostatic pressure and c_1 and c_2 are functions of the two non zero invariants of $\boldsymbol{d}$.

"It is easy to see that this is simply the particular case of the Reiner constitutive equation when the fluid is incompressible, although the point of departure I adopted was somewhat more fundamental than that of Reiner. (Reiner's point of departure was essentially that $\boldsymbol{\sigma}$ is an isotropic polynomial in $\boldsymbol{d}$, while mine was the assumption that the stress depends only on the velocity gradients and that the stress - velocity gradient relation must be unaltered by the superposition on the assumed state of flow of a rigid angular motion.)"

For the above reasons, the fluid represented by the constitutive equation (4.3) is justly known as the Reiner-Rivlin fluid.

In his paper, Rivlin (1948a) also provided solutions for rectilinear shear, torsional flow, Couette flow and later (Rivlin 1949a) Poiseuille flow in a circular tube. In so doing he essentially argued that, for the flows considered, an element of fluid was subjected to a constant shearing flow, on which was superposed a constant angular velocity, and that for this reason a constitutive equation did not need to take into account acceleration gradients and so on, the velocity gradient itself being sufficient. If this argument were valid, it would have meant that equation (4.3) could be used with generality in viscometric flows. However, the argument was incorrect, as was pointed out by Oldroyd (1950) and acknowledged by Rivlin a little later (cf. Rivlin 1971). This seemingly insignificant exchange of ideas is important, since it highlighted for the first time the important conclusion that it is not correct to assert that an elastic fluid cannot be distinguished from an (inelastic) Reiner-Rivlin fluid in steady-state experiments. There was therefore greater impetus to seek constitutive equations for *elastic* liquids, a subject which was to dominate the 1950s.

4.2. Elastic Liquids

The 1950s proved to be a rich time of creative writing in Continuum Mechanics. Indeed, the period can be viewed as a watershed between the preparatory work of previous decades and the mushrooming activity that followed the successful formulation work of Oldroyd, Rivlin, Ericksen, Green, Coleman and Noll.

In common with earlier eras of the subject, the developments were not without friction and, in retrospect, it is clear that the cult of personality plagued what should have been a period of unparalleled and unhindered progress.

In an important paper, **Oldroyd** (1950) addressed the problem of the formulation of rheological equations of state. His skilful organization enabled him in a mere twelve pages to capture and unify a whole field of science with clarity, style and economy of words. In some respects, the paper was written before its time, with other workers in the field reluctant to admit its breadth and generality. This led to rediscovery and duplication and, ultimately, unnecessary polarization. Important later developments, which should have been viewed as complementary to Oldroyd's work, were written in isolation, often without reference to Oldroyd's work. Continuum Mechanics would have been immensely enriched if the subsequent work in the middle and late 50s had been less myopic. This is especially so of the important and influential work of **Coleman** and **Noll**, the extensive papers by Noll (1955, 1958), for example, containing no reference to Oldroyd. This unfortunate state of affairs meant that by the end of the 50s, it was almost incumbent on workers in the field to identify with either Oldroyd or Noll, as if they were in deep conflict. The situation was complicated by the fact that **Truesdell**, who was recognized at the time as an important interpreter of scientific ideas, came down firmly in favour of the American work. It would seem that Truesdell viewed the work of his former student Noll (1955) as definitive and all embracing, whilst only passing (and sometimes grudging) references were made to Oldroyd's work. A few examples will suffice to illustrate the point.

In an influential General Lecture presented to a IUTAM meeting in Haifa, Israel in 1962, Truesdell (1964) addressed the problem of 'Second Order Effects in the Mechanics of Materials'. In the lecture, no reference was made to Oldroyd's work. Many felt that this was an unfortunate oversight, especially since Oldroyd was in the audience. Some of Oldroyd's frustrations were vented through an extemporary Closing Summary, a written version of which was subsequently published in the Conference Proceedings (Oldroyd 1964). The text contains a good example of Oldroyd's dry humour:

> "We may take comfort from the fact... that the sequence of physical principles and mathematical ideas that are required for a completely general discussion of equations that describe rheological behaviour need be no more complicated in, say, America in 1955 than it was in Europe in 1950".

So by 1962, philosophical divisions were clear, to the detriment of the field as a whole. The objective observer, with the benefit of hindsight, should be able to sympathize with Oldroyd's frustration. A feeling of injustice led Oldroyd to ignore, in a reciprocal fashion, much of the later work on the formulation of constitutive equations, so that he was at best defensive and at worst dismissive. Hence, there is scarcely a reference to the work of Rivlin, Ericksen, Green, Coleman and Noll in Oldroyd's subsequent papers, except for one negative comment (Oldroyd 1965) concerning the simple-fluid concept of Coleman and Noll.

Truesdell was not always so dismissive of Oldroyd's contribution as at Haifa, although the comments rarely erred on the side of effusion. For example:

"The Principle of Material Indifference... was used in some generality without a statement of what was really going on by Oldroyd in 1950. The principle was really recognized and stated explicitly by Noll in 1955" (Truesdell 1960a).

"A general solution was given in 1950 by Oldroyd, who however did not state the problem he had solved clearly enough for other students to see what it was, until they found it out for themselves. In 1955, the whole matter was set straight by Noll..." (Truesdell 1960b).

"The Oldroyd 1950 paper is the first to recognize the general programme of rational continuum mechanics and to give a prescription for carrying it through. The reader familiar with more recent theories will recognize in this paper fairly general and fairly clear statements of what are now called the principles of determinism, contiguous action, and material indifference" (Truesdell 1965a, p43).

In summary, Truesdell's writings give somewhat minimal and belated acknowledgement of the importance of Oldroyd's 1950 paper, whilst at the same time extolling the superior virtues of Noll's contributions.

Some of the other leading players were even less generous and there is hardly a reference to Oldroyd's work in the extensive writings of Coleman and Noll.[1] Rivlin (1971) on the other hand, rather belatedly perhaps, did acknowledge Oldroyd's contributions. For example, in a section headed *"Oldroyd's fundamental paper"* we have the following assessment of the work:

"It may be remarked that Oldroyd, in this paper, also gave a clear statement of the distinction between a solid and a fluid. "An incompressible liquid is not a material for which a special reference time t_0 may be supposed to exist, such that the configuration at time t_0 has permanent significance in any subsequent motion. We deduce that the equations of state can be written in a form which does not involve a strain tensor e_{ik} explicitly." This is precisely the definition of an incompressible fluid which is widely attributed to Noll (1955, 1958). We see that Oldroyd anticipated Noll, here, by at least five years."

Oldroyd was not a lone advocate of the convected-coordinate representation of constitutive modelling. It was also championed by A S Lodge (see, for example 1951) and in an Historical Note in his second book, **Lodge** (1974, p255) remarks "We take issue with the historical account given by Coleman et al (1966, p86) who state 'The first to formulate a properly invariant general theory applicable to finite deformations, however, were Green and Rivlin in 1957... . Noll employed ideas similar to those used by Green and Rivlin... Noll introduced... a general notion of fluidity, namely the concept of a simple fluid on which the present book is based." Oldroyd's (1950) paper contains a properly invariant general theory applicable to finite deformation, which includes all Noll's 'simple fluids' as Lodge and Stark (1972) have proved."

It is interesting to conjecture on the reasons for the unfortunate happenings of the 50s and early 60s in the area of constitutive equations. Was it the preoccupation in the final pages of Oldroyd's 1950 paper with simple models like those designated A and B, which left interested readers unimpressed? This is certainly suggested by such comments as

"In particular, the liquids of type (A) and (B) have nothing to recommend them except as examples" (Truesdell 1965a, p44).

"Unfortunately (in my opinion) Oldroyd based most of his subsequent work on a constitutive equation which, while it has the proper behaviour with respect to a superposed rotation and is appropriate to an isotropic fluid, is arbitrarily special in form and, moreover, does not even have the merit of being simple from the point of view of calculation"

[1]Interestingly, a recent biography of Walter Noll (Ignatieff, 1996, p29) highlights this oversight. Referring to Noll's PhD studies, it remarks "Walter Noll didn't have time to look through a large quantity of scientific literature on the subject. He escaped doing this with the following remark "Since all this literature concerns only various linearizations and 'approximations' and in generality does not go beyond Zaremba's work, it will not be considered here". In this case, it was a mistake, since he missed an important paper of J G Oldroyd, which included, in particular, something like Noll's principle of isotropy of space".

(Rivlin 1971).

This can only be a part of the answer. The most likely explanation is less scientific and more associated with personalities and human nature. If that is the case, a very fruitful period of research would have been made so much more coherent to newcomers to the field by a little objectivity on the part of the leading players. It is now generally conceded that there is room for both the convected-coordinate formalism of Oldroyd and the simple-fluid developments of Coleman and Noll et al; it is simply a matter of background and taste. It is sad that a whole generation of workers had to decide between the respective approaches, as if they were mutually contradictory.

4.3. Personalities

Rheology, like many other fields, has had more than its fair share of examples of the cult of personality. This is seen as clearly as anywhere in the area of constitutive equations. Two of the major players who have had a very influential impact on the field were Rivlin and Truesdell. In the heady days of the 50s, they were sometimes seen in tandem, giving out the same message. At that time there was apparently a large measure of mutual respect. This is nowhere better seen than on the occasion of Truesdell's Bingham Medal presentation at the Fourth International Congress on Rheology held at Brown University in 1963. In introducing Truesdell, Rivlin (1965) remarked:

"The award is made to Professor Truesdell in recognition of his profound contributions to the rational mechanics of nonlinear materials. The contributions consist not only of original papers, but of remarkable articles which are ostensibly expository, but in their breadth and depth, in their originality, erudition and historical perspective, have had a profound effect on the development of modern theoretical rheology."

Truesdell (1965b) responded:

"It is a pleasure to be allowed to make this report at a meeting organized by Mr Rivlin, whose splendid papers of 1948-55, as well as providing a starting point for most of the later work, gave the field the directness and concreteness necessary for real theory."

For reasons that are not immediately clear to a bystander, the apparent mutual respect between Rivlin and Truesdell was not sustained.[2] This is not so clear in the writings of Truesdell, except for a rather belated entry into a relatively recent controversy concerning the standing of simple-fluid theory (see, for example, Truesdell 1984, Rivlin 1984b). However, the change of attitude is abundantly clear in Rivlin's writings:

"...However, Truesdell's 1965 article with Noll *"The non-linear field theories of mechanics"* — and his voluminous later writings, impressive and admirable though they be in many respects, are seriously marred by his evident contempt for physical reasoning and insight and by a tendency to present the work of his protégés as paradigms, without regard to its originality or its physical or mathematical soundness.

"In his writings, Truesdell evidences a strong taste for the dramatic and so there has been created a fantasy world in which various savants produce a stream of principles, fundamental theorems, capital results, and work of unusual depth. No matter that,

[2]In a recent "Autobiographical Postcript" Rivlin (1996) throws some light on the change." ...I enjoyed a friendship with Clifford that lasted for more than fifteen years, but ultimately foundered, to my great regret, on his inability to accept my criticism of some of the work of his protégés".

on examination and stripped of the often irrelevant mathematical verbiage with which they are surrounded, they frequently turn out to be known results in disguise, or trivial, or physically unacceptable, or mathematically unsound, or some combination of these. Nonetheless, they have been widely and uncritically reproduced in the extensive secondary literature" (Rivlin 1984a).

The scenario in the area of constitutive equations is therefore one of strong personalities, interacting without undue warmth, during an extraordinary period of research activity. Coleman and Noll remained rather detached and disinterested in the clash of personalities, Green was far too much a gentleman to get involved, as was Ericksen. Oldroyd, for all his inherent shyness was forced into the fray, ably assisted by Lodge, although this assistance seems to have been met by a general lukewarmness on Oldroyd's part — witness his rather cool review of Lodge's second book (Oldroyd 1976).

Oldroyd remained very much a loner, the rift developed between Rivlin and Truesdell, and it was left to engineers, particularly in the USA, to bring some objectivity and sanity to this important field of research. Metzner[3] and Bird, in particular, helped to popularize the formulative ideas of Oldroyd (as well as those of Coleman and Noll), and commented favourably on Oldroyd's use of simple models to illustrate the general theory. At the same time, important books were written by Lodge (1964, 1974), Bird et al (1977a,b) and Schowalter[4] (1978).

The dust has now settled, and the hype recognized for what it was. The cult of personality simply delayed rather than prevented developments. Nowadays, it is possible to appreciate the massive contributions of all the leading players, without the need to take sides.

4.4. Progress is Made

The uneven and often confused developments referred to in §4.2 should not be allowed to mask the important advances in the field of constitutive equations which took place in the 50s and 60s, beginning with the work of Rivlin and Ericksen (1955).

The sequence of events which led to the introduction of the so called Rivlin-Ericksen fluid are conveniently described by Rivlin (1971) himself: "I came to the Naval Research

[3]Arthur B Metzner was born in Gravelbourg, Saskatchewan, Canada on the 13th April 1927. He received a BSc in Chemical Engineering from the University of Alberta in 1948 and an ScD from M. I. T. in 1951. His move to the University of Delaware in 1953 coincided with its emergence as an international centre of excellence in rheology. He has received many honours during a distinguished career, including an Honorary Degree from the Katholieke Universiteit, Leuven, Belgium in 1975, the award of the Bingham medal in 1977 and election to the National Academy of Engineering in 1979. He formally retired as the H Fletcher Brown Professor in 1993, but he has continued to serve the Society of Rheology in various capacities. His successful tenure as Editor of the Journal of Rheology came to an end in 1995, and, in 1997, he received the S.O.R. Distinguished Service Award.

[4]William R Schowalter was born in Milwaukee, Wisconsin on the 15th December 1929. He received a BS in Chemical Engineering from the University of Wisconsin, Madison in 1951 and an MS and PhD from the University of Illinois at Urbana-Champaign (UIUC) in 1953 and 1957, respectively. For most of his distinguished academic career, a period covering 31 years, he was associated with Princeton University, where he served as Chairman of the Department of Chemical Engineering and Associate Dean of the School of Engineering and Applied Science. He is now Dean of the College of Engineering at UIUC. Schowalter was elected to the National Academy of Engineering in 1982 and he received the Bingham medal in 1988.

Laboratory in April 1952 and was asked to head a small group which included Ericksen and Toupin. Very shortly after my arrival, I embarked on the construction of a continuum mechanical theory in collaboration with Ericksen. This theory was essentially complete when I left the Naval Research Laboratory to join the Faculty of Brown University in August 1953, having decided to remain in the United States. However, for a variety of reasons the work was not published until 1955".

"We made the assumption that, in a viscoelastic material, whether fluid or solid, the Cauchy stress depends on the deformation gradients, the velocity gradients, the acceleration gradients, the second acceleration gradients and so on... . We then introduced the condition that the superposition on the assumed deformation of a rigid time-dependent rotation causes the stress to be rotated correspondingly. We also introduced the assumption that the material is isotropic. (Unfortunately, we did not separate these two assumptions. This led to no error, but the derivations would undoubtedly have had greater clarity had we done so)."

In the case when the material is an incompressible isotropic fluid, the resulting constitutive equation up to order n can be written in the form:

$$\boldsymbol{T} = \boldsymbol{f}(\boldsymbol{A}_1, \boldsymbol{A}_2, ...\boldsymbol{A}_n), \tag{4.4}$$

where $\boldsymbol{f}$ is an isotropic matrix function of its arguments, which are the kinematic tensors that have since become known as the Rivlin-Ericksen tensors $\boldsymbol{A}_n$. $\boldsymbol{A}_1$ is just $2\boldsymbol{d}$ and the higher-order tensors are essentially the same (apart from a factor of 2) as the higher rates-of-strain $e_{ik}^{(n)}$ derived by Oldroyd in his 1950 paper, a fact which does not appear to be generally appreciated.

Obtaining restrictions on the form of $\boldsymbol{f}$ involved an extensive research programme in invariant theory. This was carried out by Rivlin, ably assisted by A J M Spencer and G F Smith. Of major importance was a discovery made by Rivlin (1955) himself that, when $n = 2$, $\boldsymbol{f}$ can be written in the form:

$$\begin{aligned} \boldsymbol{f} \; = \; & \alpha_1\boldsymbol{A}_1 + \alpha_2\boldsymbol{A}_2 + \alpha_3\boldsymbol{A}_1^2 + \alpha_4\boldsymbol{A}_2^2 \\ & + \alpha_5(\boldsymbol{A}_1\boldsymbol{A}_2 + \boldsymbol{A}_2\boldsymbol{A}_1) + \alpha_6(\boldsymbol{A}_1^2\boldsymbol{A}_2 + \boldsymbol{A}_2\boldsymbol{A}_1^2) \\ & + \alpha_7(\boldsymbol{A}_1\boldsymbol{A}_2^2 + \boldsymbol{A}_2^2\boldsymbol{A}_1) + \alpha_8(\boldsymbol{A}_1^2\boldsymbol{A}_2^2 + \boldsymbol{A}_2^2\boldsymbol{A}_1^2), \end{aligned} \tag{4.5}$$

where the α's are functions of

$$\begin{aligned} & tr\boldsymbol{A}_1^2,\ tr\boldsymbol{A}_1^3,\ tr\boldsymbol{A}_2^2,\ tr\boldsymbol{A}_2^3, \\ & tr\boldsymbol{A}_1\boldsymbol{A}_2,\ tr\boldsymbol{A}_1^2\boldsymbol{A}_2,\ tr\boldsymbol{A}_1\boldsymbol{A}_2^2,\ tr\boldsymbol{A}_1^2\boldsymbol{A}_2^2. \end{aligned} \tag{4.6}$$

Since, for viscometric flows, $\boldsymbol{A}_n = \boldsymbol{0}$ for $n > 2$, Rivlin was able to utilize (4.5) to extend his earlier viscometric-flow analysis for the Reiner-Rivlin fluid to the general Rivlin-Ericksen fluid.

Markovitz and Williamson (1957) used Rivlin and Ericksen's 1955 theory to give a framework for their viscometric experiments and they rewrote (4.5) as

$$\boldsymbol{T} = \alpha_1^1\boldsymbol{A}_1 + \alpha_2^1\boldsymbol{A}_2 + \alpha_3^1\boldsymbol{A}_1^2, \tag{4.7}$$

where α_1^1, α_2^1 and α_3^1 were functions of the square of the shear rate. Ericksen, in two papers (Criminale et al 1958, Ericksen 1960), also saw that, for shearing motions, or viscometric flows as they came to be called, the Rivlin-Ericksen equations would reduce to the form (4.7). The 1960 paper was the written record of a lecture given in late April 1958. Ericksen worked out the stresses in terms of the α's and stated "These relations say merely what is sometimes granted as obvious on symmetry grounds, i.e., that the normal stress differences... as well as the apparent viscosity, ..., should be even... functions of (the shear rate), while the (other) shear stresses... should vanish". The Criminale-Ericksen-Filbey (CEF) constitutive equation (4.7) has subsequently been used a great deal beyond its strict domain of relevance, occasionally with disastrous results.

The next logical step in the development of the general theory was undertaken independently by Green,[5] Rivlin and Spencer (1957, 1959, 1960) and Coleman and Noll (see, for example, Coleman and Noll 1960, 1961). So, for incompressible fluids, for example, attention moved to the so called 'simple fluid',[6] with the extra stress tensor **T** related to a suitable measure of deformation $\boldsymbol{C}$ (the Cauchy-Green tensor) through a functional equation of the form:

$$\boldsymbol{T} = \mathop{\boldsymbol{F}}_{-\infty}^{t} [\boldsymbol{C}(t')]. \tag{4.8}$$

Equation (4.8) means that the stress at a particle at the present time ($\boldsymbol{T}(t)$) depends on the entire past history of the local deformation ($\boldsymbol{C}(t')$) measured relative to the present time.

The Green-Rivlin theory involved the replacement of the functional in (4.8) by a series of repeated integrals. The Coleman and Noll development was more elegant and essentially relied on the fact that different measures of fading memory are most easily and naturally defined in terms of topologies in function spaces, and that the essence of the matter lies in the amount of smoothness presumed of the constitutive functional in such topologies (cf. Truesdell 1964).

In a fruitful period of research, Coleman and Noll set about investigating some of the implications of simple-fluid theory. An influential paper (Coleman and Noll 1960), on the 'slow flow' of simple fluids with fading memory, led to a series of approximations, which have since become known as the hierarchy equations of Coleman and Noll. The first-order approximation is the Newtonian limit, while the second-order approximation leads to the

[5] A E Green was born in London on 11th November 1912. He studied Mathematics at Cambridge University and graduated in 1934 with a First Class Honours degree. He then carried out research under the direction of Sir Geoffrey Taylor, obtaining a PhD degree in 1937. From 1936 to 1939, Green held a Research Fellowship at Jesus College, Cambridge. He then moved to Durham University to take up a lecturing appointment. In 1948, he became Professor of Applied Mathematics at Kings College, Newcastle upon Tyne, which at that time was part of the University of Durham. He remained there until 1968, when he accepted the Sedleian Chair of Natural Philosophy at Oxford University, a post he filled until his retirement in 1977. His research continued to flourish after 1977 and the list of publications contained in a book compiled in his honour (Naghdi et al 1994) highlights thirty or so research papers published since his formal retirement. He was elected to a Fellowship of the Royal Society in 1958.

[6] In some respects, the functional equation (4.8) is anything but 'simple'. Noll's reason for choosing the term is summarized in this private communication to one of the authors (KW). "The dictionary gives at least two meanings for *simple*: One is that of one-fold, having only one ingredient. Another is that of being not complicated. I meant the first meaning, *simple* as an abbreviation for *of grade (order) one*. I remember that Hershel Markovitz thought at the time that the term would not find acceptance because of the second meaning of *simple*".

more useful second-order fluid:

$$\boldsymbol{T} = \alpha_1 \boldsymbol{A}_1 + \alpha_2 \boldsymbol{A}_2 + \alpha_3 \boldsymbol{A}_1^2, \tag{4.9}$$

where the α_is are constants.

To say that the Coleman and Noll paper was somewhat taxing to those who were not mathematically minded would be an understatement, but Rivlin (1984a) has been quick to reassure such people that the hierarchy equations can be viewed as truncations of the Rivlin-Ericksen constitutive equation.

Another important Coleman and Noll (1961) contribution concerned the limit of 'small deformation' which led to a series of integral approximations, the most important being that associated with so called 'finite linear viscoelasticity', with equations

$$\boldsymbol{T}(t) = \int_{-\infty}^{t} M_1(t-t')\boldsymbol{C}(t')dt'. \tag{4.10}$$

The paper contained a useful discussion on the relation between this genuinely non linear theory and the usual equations of linear viscoelasticity.

The work of A C Pipkin[7] should be mentioned at this point. His work on constitutive constraints was invariably clear and concise and often led to helpful simplifications. For example, in the case of the integral expansions arising out of either the work of Coleman and Noll (1961) or Green and Rivlin (1957,1960) and Green et al (1959), he obtained simpler expressions by removing unnecessary clutter (see, for example, Pipkin 1964).

A consistent use of the approximations embodied in (4.9) and (4.10) allowed Coleman and Markovitz (1964a, b) to derive helpful constraints on some of the parameters. Of particular importance is the result

$$\alpha_2 \leq 0. \tag{4.11}$$

Although the Coleman and Markovitz papers have received nothing like the same publicity as some of the papers in the series, the constraint (4.11) is an important one. Sadly, the constraint has sometimes been ignored (since it appears to contradict certain results coming from 'thermodynamics'). This has resulted in a substantial literature, which, although it may have kept applied mathematicians active and in employment, has little relevance to the real world of experiments (cf. the related comments of Giesekus 1989).

4.5. Other Relatively Simple Equations

In one sense, the general functional equation (4.8), suitably embellished with an appropriate function space and norm, can be viewed as having sufficient generality to encompass most, if not all, isotropic elastic liquids. Its major drawback is of course its inability to make specific predictions for anything but the simplest flow fields.

Many materials of industrial importance can be classified as elastic liquids and the growing pressures to handle such liquids in a process-modelling context has presented

[7] A C (Jack) Pipkin was born in 1931. He received a BS degree from the Massachusetts Institute of Technology in 1952 and a PhD in mathematics from Brown University in 1959. Following a year as Research Associate in the Institute for Fluid Dynamics and Applied Mathematics at the University of Maryland, he joined the Faculty at Brown University, where he spent the remainder of his distinguished academic career. He died suddenly in 1994.

rheologists with significant challenges over several decades. The need for 'answers' led to a search for simple (approximate) constitutive equations, which had predictive value at least in a qualitative or semi-quantitative sense. The need for such pragmatism was not universally welcomed or accepted by many theoreticians, and some simply refused to move outside the haven of generality. For others, pragmatism prevailed and notable advances were made. These were not only driven by process-modelling considerations, but arose out of more esoteric considerations, usually emanating from a desire to ensure that the constitutive equations reflected the physics of the fluid microstructure.

These developments have resulted in the appearance of a plethora of constitutive equations, many of which have had their faithful adherents and supporters, although it must be said that the popularity of a given constitutive model has often been ephemeral.

One strand of activity can be traced back to the pioneering work of Fröhlich and Sack (1946) for a very dilute suspension of elastic spheres in a viscous medium and of Oldroyd (1953) for the related problem of a dilute emulsion consisting of two Newtonian constituents. Working within a linearity constraint, they showed that the appropriate constitutive law could be expressed in the linear form:

$$\boldsymbol{T} + \lambda_1 \frac{\partial}{\partial t}\boldsymbol{T} = 2\eta_0 \left[\boldsymbol{d} + \lambda_2 \frac{\partial}{\partial t}\boldsymbol{d}\right], \tag{4.12}$$

with the material constants η_0, λ_1 and λ_2 related to the properties of the individual constituents and to the concentration.

Oldroyd certainly saw the availability of equations like (4.12) as a vital part of the basic Scientific Method, and he used the formulation principles developed in his 1950 and 1958 papers to construct constitutive equations, valid for all conditions of motion and stress, which reduced to (4.12) under the linearity constraint; the hope being that such equations would provide an adequate description of observed behaviour over a wide range of general flow conditions.

One of the simplest and most popular of the equations thus constructed was designated Liquid B by Oldroyd (1950). This can be written

$$\boldsymbol{T} + \lambda_1 \overset{\nabla}{\boldsymbol{T}} = 2\eta_0 \left[\boldsymbol{d} + \lambda_2 \overset{\nabla}{\boldsymbol{d}}\right], \tag{4.13}$$

where ∇ denotes the so called upper convected time derivative. The special case $\lambda_2 = 0$ is now universally referred to as the Upper Convected Maxwell (or UCM) model, the motivation for the terminology being the fact that the linear version of (4.13) (with $\lambda_2 = 0$) was that proposed by James Clerk Maxwell in 1867/68 (see Chapter 1).

The Oldroyd B and UCM models have had a popularity far beyond expectation and anticipation. The reasons for this are related to a number of factors. Their *relative* simplicity as generally-valid constitutive equations has obviously been an attraction, especially in modern fields like the Numerical Simulation of Viscoelastic Flow (see Chapter 8), where simple models have been very helpful (if not essential) in the development of suitable numerical strategies.

The experimental advent of the so called **Boger** fluids (see, for example, Boger 1977/78)[8] also helped to broaden interest in models like (4.13). To a good approximation, in a steady

[8]Walters (1979) is responsible for the term 'Boger fluid'. See also Choplin et al (1983) and Binnington and Boger (1985).

simple shear flow, these fluids (very dilute solutions of a high molecular weight polymer in a very viscous solvent) have a constant viscosity and a quadratic first normal stress difference over a reasonable range of shear rates, and for many years the Oldroyd B model was seen as a useful first approximation for such fluids, since the viscometric functions for this model are broadly consistent with experiment. However, as the field developed, so more complicated models than (4.13) were found to be necessary, but there is no doubt that the combination of a new field (involving computer simulation) and a new (Boger) fluid has given unusual prominence to a model which was introduced nearly 50 years ago.

Notwithstanding the special circumstances surrounding Boger fluids and the Oldroyd B model, it quickly became evident that satisfying the dual constraints of tractability and adequate material description was not easy for rheologically complex fluids. Depending on the definition of 'adequate', some would go as far as to say that the task was impossible.

However, such pessimism did not prevent some worthy attempts, typified perhaps by the work of White[9] and Metzner (1963), who proposed an equation of the form:

$$\boldsymbol{T} + \lambda(I_2) \overset{\nabla}{\boldsymbol{T}} = 2\eta(I_2)\boldsymbol{d}, \tag{4.14}$$

where I_2 is the second invariant of $\boldsymbol{d}$. By a suitable choice of the functions η and λ, the viscosity and first normal stress difference in a steady simple shear flow can be simulated exactly. At the same time, the so called White-Metzner model suffers from the limitations of having just one relaxation time, a zero second normal stress difference and an extensional viscosity which becomes infinite at a finite value of the extensional strain rate. In many ways, this is a good example of the difficulties of meeting the dual constraints of relative simplicity and adequate prediction, and, as one would expect, the ad hoc nature of this process of model building has not been to everyone's taste. This criticism could not be levelled at the many attempts to seek constitutive equations based on microrheological ideas (see Chapter 5). Most of these essentially bypassed the linear-viscoelasticity stage inherent in the philosophy of Oldroyd, an early example of this being the Lodge (1956) rubber-like liquid. Here, the microstructure for a concentrated polymer solution or melt was approached through the molecular theory of Gaussian networks, continuous deformation being allowed through the creation and destruction of temporary network junctions. Affine motion was also assumed in that the network junctions were assumed to move as points of the equivalent continuum. In this way, the following integral constitutive equation was derived:

$$\boldsymbol{T} = \int_{-\infty}^{t} \Phi(t - t')\boldsymbol{C}^{-1}(t')dt', \tag{4.15}$$

where $\boldsymbol{C}^{-1}$ is the Finger tensor. This equation is essentially a generalization of the UCM model for a distribution of relaxation times.

[9]James Lindsay White was born in Brooklyn, New York in 1938. He obtained his initial university education at the Polytechnic Institute of Brooklyn, before moving to the University of Delaware, where he received MS and PhD degrees. He spent four years in industry working for Uniroyal, and then had a productive association with the University of Tennessee, during a stay of sixteen years. In 1982, he became Director of the Center for Polymer Engineering at Akron University. White was the founding President of the Polymer Processing Society from 1985 to 1987, and he is still Editor-in Chief of the Society's Journal. He was awarded the S.O.R. Bingham medal in 1981.

Lodge's work on the rubber-like liquid was influential in motivating several later developments. For example, the affine motion assumption was relaxed in the work of Johnson and Segalman (1977) and Phan-Thien[10] and Tanner (1977). For the special case of one relaxation time, the resulting analyses led to constitutive equations which can be written in the tensorial form:

$$f(\epsilon T_{mm}\lambda/\eta_o)T_{ik} + \lambda \overset{\square}{T}_{ik} = 2\eta d_{ik}, \tag{4.16}$$

where the time derivative denoted by $\square$ is given by

$$\overset{\square}{T}_{ik} = \left(1 - \frac{\alpha}{2}\right) \overset{\nabla}{T}_{ik} + \frac{\alpha}{2} \overset{\Delta}{T}_{ik}, \tag{4.17}$$

Δ being the lower convected time derivative. The function f is an increasing function of $\epsilon T_{mm}\lambda/\eta_o$, where ϵ is usually a small parameter.

In the Johnson and Segalman work $\epsilon = 0$, and, in most subsequent applications, α has been taken to be a constant or often zero.

The Johnson/Segalman (1977) and Phan-Thien/Tanner (1977) papers were important developments in that the constant viscosity restriction in the Lodge rubber-like liquid was removed by a suitable choice of the constant α. The same constant allowed for a non-zero second-normal-stress difference and the ϵ in the Phan-Thien/Tanner model permitted a more realistic extensional-viscosity response. The equation of Giesekus (1982, 1985) also possesses many positive features.

The relative simplicity (and attendant inadequacies) of the Lodge rubber-like liquid constitutive equations were also obvious to other workers in the field and it was inevitable that more complicated integral models would be proposed. An influential model, which has since become known as the KBKZ model, was developed independently by Kaye (1962) working in the UK and by Bernstein, Kearsley and Zapas (1963) working at the National Bureau of Standards in Washington. The resulting equation can be written in the form:

$$\boldsymbol{T} = \int_{-\infty}^{t} \left[\frac{\partial\Omega}{\partial I_1}\boldsymbol{C} - \frac{\partial\Omega}{\partial I_2}\boldsymbol{C}^{-1}\right] dt', \tag{4.18}$$

where Ω is a potential function and I_1 and I_2 are (for incompressible fluids) the two non-zero invariants of the deformation tensor $\boldsymbol{C}$. Kaye's work was a straight-forward attempt to generalize Lodge's (1956) ideas, using the stress-strain relations for an incompressible elastic solid as the basic starting point. No attempt was made to provide a detailed physical interpretation. Unfortunately, the work was not published in prominent journals and it lay hidden for many years. It is now acknowledged to be a useful contribution to

[10]Nhan Phan-Thien was born in An Giang, South Vietnam in 1952. His academic career has been concentrated almost exclusively at the University of Sydney, and he is now an Australian citizen. He obtained his PhD in 1978 and was awarded a Personal Chair in Mechanical Engineering in 1991. Phan-Thien has written two books, one in 1994 with Professor S Kim of U. W. Madison, entitled "Microstructures in elastic media: Principles and computational methods" and one in 1997 with Dr R R Huilgol of Flinders University entitled "Fluid mechanics of viscoelasticity". He was awarded the Australian Society of Rheology Medal in 1997.

the evolving work on constitutive modelling, and the equation (4.18) is today known as the KBKZ equation, the first K recognizing Kaye's contribution.

Bernstein, **Kearsley** and **Zapas** attempted to provide a physical justification for the model by appeal to 'thermodynamics', but such an interpretation was not met with universal acclaim. For example, J L White and N Tokita (1967) did not use an elastic potential representation and proposed a model of the form:

$$\boldsymbol{T} = \int_{-\infty}^{t} \left[\Psi_1 \boldsymbol{C} + \Psi_2 \boldsymbol{C}^{-1}\right] dt'. \tag{4.19}$$

Rivlin and Sawyers (1971) also proposed (4.19) and said that they considered the KBKZ potential neither "well-based nor necessary". The form (4.19) has not demonstrated great advantages over the KBKZ model, but on the other hand no difficulties due to thermodynamic unsoundness has, as far as we know, been found with (4.19); such an example could probably be constructed.

Almost universally, the kernel functions in both (4.18) and (4.19) are split into separable time-dependent and strain-dependent contributions, basically for reasons of tractability. Although some limited experimental evidence seemed to be in sympathy with the proposed splitting, it continues to be questioned.

Further valuable developments of the single integral constitutive models were introduced by Wagner[11] (see, for example, Wagner 1978), involving reversible and irreversible damping functions, in order to simulate observed behaviour in a more realistic manner. One of the problems that became noticeable with the KBKZ model and many other equations was the 'recoil problem'. If a tensile sample were deformed suddenly, held for a while, and then freed of stress, it was found that the observed strain recovery was much less than the predicted recovery. Thus, most equations are too "elastic" in nature in such reversing deformations. Wagner, in 1978, and in many papers since then, used an "irreversibility" argument based on microstructural theory to partially rectify these shortcomings; this quest continues (Larson 1988).

So, as the field developed, constitutive equations were introduced which had some input from basic continuum modelling, some from microrheology, some from the need to simulate real experimental data and a combination of all three. All the time, the dual requirements of tractability and adequate flow prediction were elusive.

A major event took place in the late seventies with the outworking of the 'reptation' model by **Doi** and **Edwards** (see, for example, Doi and Edwards 1986).[12] This had a significant impact on the field and led to many important developments. In its general

[11]Manfred H Wagner was born in Stuttgart, Germany on 9th September 1948. As a Physics major, he took a PhD in Chemical Engineering at the Institute for Polymer Processing in Stuttgart. There followed two years post-doctoral study with Professor J Meissner and then nine years of industrial experience in the carbon and graphite field, before he returned to Stuttgart as Professor of Fluid Dynamics and Rheology. Manfred Wagner has been President of the Germany Society of Rheology since 1991. He is currently Secretary of the European Society of Rheology and a member of the Executive Committee for Rheology 2000.

[12]A related development associated with the names of Curtiss and Bird took a somewhat different approach. This work is covered in Part 2 of 'the Dynamics of Polymeric Liquids' books co-authored by Bird, Curtiss, Armstrong and Hassager (1987b) (see §5.5).

form, the Doi-Edwards theory has no closed-form constitutive equation. When added conditions like the 'independent alignment' assumption have been incorporated, the Doi-Edwards model then takes on a familiar integral form, but the resulting predictions do not match up at all well to experiment (see, for example, Marrucci 1986, Marrucci and Grizzuti 1986) and such excursions to seek recognizable constitutive equations in the Doi-Edwards development have not always been successful and, in fact, have sometimes been counterproductive. At the same time, the basic Doi-Edwards ideas and research have been immensely influential and are attractive as a physical picture of polymers (see also Chapter 5).

As the work on modelling for concentrated polymer solutions and polymer melts was taking place, related developments were underway for dilute polymer solutions. The bead-spring models were extended to accommodate more realistic spring laws, culminating in the FENE (Finitely Extensible Non-Linear Elastic) dumbbell models. The so-called Chilcott-Rallison (1988) model is an example of later developments (see also Chapter 5).

In this sketch of developments in constitutive modelling, we have concentrated on the major players and advances. Important contributions have also been made by **Marrucci** (see, for example Acierno et al 1976), Giesekus (1982) and Leonov (1987). These and the workers already referred to have clearly been guided by a combination of the following constraints:

1. The equations must satisfy the basic formulation principles.

2. The equations should reflect the microstructure. The Lodge rubber-like liquid, and the Doi-Edwards model are good examples of this philosophy.

3. The equations should lead to simulated behaviour in simple flows, which matches that found for real liquids. The White-Metzner work is a good example of an attempt to satisfy the constraint, without any specific reference to (2).

4.6. Overview

A review of the relevant literature makes it abundantly clear that the search for acceptable constitutive equations for rheologically complex fluids has not been straightforward and the historical development has been patchy. At the same time, it has been possible to highlight specific thrusts, which have been influential in the developing story.

Oldroyd's basic formulation work is now viewed as seminal, but his early hope and expectation that relatively simple constitutive equations would suffice for complex fluids like polymer solutions and emulsions is now, with the benefit of hindsight, viewed as over optimistic.

The concentration on 'generality' in the work of Coleman and Noll and others in the late 50s and 60s is also viewed as important and revealing. At the same time, the implied deprecation of any move away from the haven of generality has not been followed by most workers in the field. If it had, research into theoretical rheology would have come to a premature halt in the mid sixties.

The field was left with a vacuum, which it is still trying to fill. Significant advances have been made in many quarters by a judicious combination of a correct use of the formulation principles, a suitable mathematical form for the microrheology, and straightforward pragmatism.

4.7. Anisotropic Fluids

The innovative and creative work on constitutive modelling for *isotropic* non-Newtonian fluids, which so characterized the 50s and early 60s, was complemented by important developments in the study of *anisotropic* fluids. The Swedish scientist Oseen (1925) had prepared the ground much earlier, but his work was not particularly accessible, being published in German in a Swedish astrophysical journal. However, we believe that J L Ericksen[13] is seen as the father of modern developments in anisotropic fluid theory and it is his name, along with that of F M Leslie, which is now associated with the most influential theory.

Leslie, who had begun his research career studying the flow behaviour of isotropic fluids of the Oldroyd type (see, for example, Leslie 1961), was introduced to the distinctive challenges of anisotropic-fluid theory by A E Green, his then Head of Department at the University of Newcastle. There followed a year's sabbatical leave with Ericksen at the Johns Hopkins University (1966-67) and this was to set the course of Leslie's future career.

R J Atkin and P K Currie were to follow Leslie to Johns Hopkins in 1967 and 1968, respectively, and they also made contributions to the evolving field.

Developments in anisotropic fluid theory were given a special boost by obvious applications to liquid crystal displays and similar devices, and in the 80s a further impetus was provided by the growth of interest in liquid crystal polymers (LCPs) like Kevlar.

Continuum theory for anisotropic fluids requires an obvious additional variable over and above those employed in the isotropic theory. This unit vector field, usually denoted by n_i, is commonly referred to as the director. Another added complication is that the balance laws of both linear *and* angular momentum play a role in the fluid dynamics of anisotropic fluids.

In view of these complications, it is not surprising that initial developments of the theory were quite special and nothing like as general as, say, the simple fluid theory being introduced by Coleman and Noll at around the same time for isotropic fluids. So, for example, the immensely influential Leslie (1968)/Ericksen (1961) theory is even in n_i, and linear in the rate of strain d_{ik} and n_i, where

$$N_i = \frac{Dn_i}{Dt} - \omega_{ip}n_p, \tag{4.20}$$

ω_{ik} being the vorticity and D/Dt denoting the material time derivative.

Under these restrictive conditions, the extra stress tensor T_{ik} can be written in the

[13]J L Ericksen was born in Portland, Oregon on 20th December 1924. Between 1943 and 1946, he was attached to the US Naval Reserve and saw active service in the Pacific. In 1947, he obtained a BS degree at the University of Washington and, in 1949, he was awarded an MA degree by Oregon State University. His PhD degree was obtained from Indiana University in 1951. From 1951 to 1957, Ericksen worked at the US Naval Research Laboratory in Washington. He moved to Johns Hopkins University in 1957 and remained there until his move to the University of Minnesota in 1983. He retired in 1990.

form:

$$T_{ik} = \alpha_1 d_{j\ell} n_j n_\ell n_i n_k + \alpha_2 N_i n_k + \alpha_3 N_k n_i + \alpha_4 d_{ik} + \alpha_5 d_{ij} n_j n_k + \alpha_6 d_{kj} n_j n_i, \tag{4.21}$$

where here the α's are the so called Leslie coefficients. Parodi (1970), applying Onsager relationships, showed that the constants can be related through

$$\alpha_6 - \alpha_5 = \alpha_3 + \alpha_2. \tag{4.22}$$

It is reasonable to enquire why such a special theory is still very much in vogue, although similarly restrictive theories for isotropic fluids have often been maligned. The answer is interesting and informative. There is no doubt that the initial motivation was pragmatic, given the obvious difficulties of attempting any level of generality. However, some of the implications of the theory, like the scaling laws for simple shear and Poiseuille flow (see for example, Leslie 1987), were shown to be in surprisingly good agreement with experiment, and any further complication to the theory would be likely to destroy that agreement. With hindsight, therefore, there would have been little merit in attempting a more sophisticated theory. For this reason, the Leslie-Ericksen theory has been and continues to be a focal point in modern developments in liquid-crystal theory. For example, between 1968 and 1974, the Nobel Laureate, P G Gennes and his group applied the Leslie-Ericksen theory to a number of experimental situations (see, for example, de Gennes 1974).

4.8. Ronald S Rivlin

Ronald Samuel Rivlin was born in London, England, on May 6th, 1915. In 1933, he went up to the University of Cambridge, where he studied Mathematics and Physics at St Johns College. By his own admission (Rivlin 1996), he did not do quite as well as expected, something he attributes in part to the time spent 'cultivating interests in a variety of cultural fields, in discussion, and in social development'. He was awarded a BA in 1937, an MA in 1939 and later in 1952 an ScD (higher doctorate).

Rivlin's father died in January 1937 and this provided a motivation to leave University and seek employment. Accordingly, he joined the General Electric Company in 1937 and remained there until 1942. During this time, he came into contact with L R G Treloar, a contact that would help mould his future research interests.

In 1942, Rivlin left GEC for the Telecommunications Research Establishment (TRE), 'to get closer to the war effort'. He stayed at TRE until 1944, when he joined the British Rubber Producers' Research Association (BRPRA) mainly at Treloar's instigation. Apart from a year spent in the USA (1946-47), he remained at BRPRA until 1952, producing some of his most influential research in the area of finite-strain elasticity. During this time, he also developed an interest in viscoelastic liquids, partly as a result of a contact with Karl Weissenberg, who showed Rivlin his rod-climbing experiments (see §7.2). This motivated Rivlin's work on the constitutive equations for what is now known as the Reiner-Rivlin fluid.

The 1946-47 visit to the US took Rivlin to the National Bureau of Standards in Washington, DC, with a part time commitment at the Mellon Institute in Pittsburg. The latter connection was particularly important, since it was there that he met his future wife, Violet, who was at the time employed at the Institute as a research chemist. They were married in June 1948.

Rivlin was granted another year's leave of absence from BRPRA in 1952 and this took him to the Naval Research Laboratory (NRL) in Washington, DC. At this time, he and his wife took the major decision to move permanently to the US and he took up an appointment as Professor of Applied Mathematics at Brown University, Providence, RI, in August 1953. Later in 1955, Rivlin took out American citizenship.

The years following 1953 saw the Division of Applied Mathematics at Brown enjoying an impressive international reputation, due in no small measure to Rivlin's presence, although he was far from being the only celebrity on the faculty at the time. Rivlin received the Bingham medal of the Society of Rheology in 1958 and later, in 1963, Brown University hosted the 4th International Congress on Rheology, with Rivlin acting as co-chairman.

The mid fifties turned out to be another very productive period for Rivlin's research and he published several influential papers on constitutive equations for viscoelastic materials, with the collaboration of a number of well known coworkers, notably J L Ericksen, A E Green and A J M Spencer.

Not so successful was Rivlin's tenure as Division Chairman, which commenced in 1958. His full and interesting 'Autobiographical Postscript' (Rivlin 1996) contains a graphic account of the tensions of the period and the pains they caused. He relinquished the chairmanship in 1963.

In 1967, Rivlin moved to Lehigh University in Bethlehem, Pennsylvania, to head the newly created 'Centre for the Application of Mathematics'. He remained in that position until his formal retirement in 1980.

Rivlin has always regarded mechanics as a branch of physics and he has often shown impatience with those who would prefer to see the subject as a branch of pure mathematics. Rivlin himself feels that this has alienated him from mathematical rheologists, and this is probably true. But the general field has undoubtedly been the richer for Rivlin's interdisciplinary approach and genuine desire to make theoretical rheology palatable to non-experts.

In his prime, at scientific meetings, Rivlin would appear as an avuncular, convivial, chain-smoking, larger than life, figure, seeking to capture the attention of a friend or colleague with a quizzical glance. By any yardstick, he must be judged as one of the leading post-war rheologists.

4.9. James G Oldroyd

James Gardner Oldroyd was born in Bradford, England, in April 1921. He attended the local Grammar School, where he matriculated in 9 subjects at the age of 14. Due to his young age, he was required to spend four years in the 6th form, where he excelled in mathematics and also became proficient in the German and French languages. He was awarded a State Scholarship and went up to Trinity College, Cambridge, in October 1939. There followed a distinguished undergraduate career; he was awarded the Rouse Ball prize in 1941 and the Mayhew prize in 1942.

Jim Oldroyd's professional career began at the Ministry of Supply Facility at Aberporth, a small village to the south of Aberystwyth on the West Wales coast. He was there from 1942 to 1945 undertaking rocket research.

After the Second World War, in 1945, Oldroyd moved to Courtaulds Fundamental Research Laboratory in Maidenhead, England. There, he had a fruitful interaction with D J Strawbridge and B A Toms, two experimenters who were involved at the time in some innovative rheometry (see, for example, Oldroyd et al 1951); Toms was also engaged in research into the 'drag reduction' phenomenon (see §7.7).

At Courtaulds, Oldroyd produced a series of remarkable papers, culminating in the 1950 paper in the Proceedings of the Royal Society on the formulation of rheological equations of state. Amongst his Courtaulds' publications is also a series in the Proceedings of the Cambridge Philosophical Society (which formed the basis of his Cambridge PhD thesis) on Bingham and non-Bingham plastic solids; these demonstrate his originality in problem solving and his considerable mathematical skills (see, for example, Bird 1988). His interest in microrheology is evidenced by his 50s work on emulsions (see, for example, Oldroyd 1953).

In 1953, Oldroyd acted as Treasurer of the 2nd International Congress on Rheology held at Oxford, just as he was leaving Courtaulds to take up the Chair of Applied Mathematics at the University College of Swansea, a constituent college of the University of Wales, at the early age of 32. In 1965, Oldroyd moved to a similar position at the University of Liverpool and remained there until his untimely death in November, 1982; he died suddenly and without warning on his way to his university office.

In 1946, Oldroyd married Marged Katryn Evans, someone he had met during his Aberporth sojourn. They had three sons.

During an illustrious career, Oldroyd won many honours. In 1964, he was awarded the Adams Prize by the University of Cambridge and, in 1980, he received a Gold Medal from the British Society of Rheology. His influence on many post-war developments in rheology was considerable; evidence of this is provided by R B Bird's fitting tribute at the opening lecture of the 1988 International Congress in Sydney (Bird 1988); his lecture was entitled 'The Two JG's' (J G Oldroyd and J G Kirkwood).

Jim Oldroyd was at first meeting quiet and reserved, but this innate shyness did not prevent him from occasionally expressing his views in a forthright fashion. He was an accomplished after dinner speaker and possessed a keen, subtle, sense of humour.

4.10. Coleman and Noll

B D Coleman

W Noll

In many ways, the major contributions of B D Coleman and W Noll to rheology are inseparable and it is convenient to consider the two scientists together. Their active collaboration essentially stretched from 1958 to 1966 and involved one book (written in collaboration with H Markovitz (Coleman et al 1966)) and over twenty research papers. It was a fruitful partnership in an important period in the development of rheological ideas and it had an immense influence, particularly amongst workers in theoretical continuum mechanics; impact on *experimental* rheologists was also significant. They are best known for their work on 'simple fluids with fading memory'.

Walter Noll is often seen as the dominant partner. He was born in Berlin on January 7th, 1925. His father, Franz Noll, originated from a Dutch family in Rotterdam.

Noll lived in Germany through the Second World War and, in 1944, he was drafted into the Air Force Signal Corps, his main duties involving air traffic control, aircraft warning, the guidance of interceptors and radio reconnaissance. In February 1945, Noll was sent to the Russian Front, but he got attached to soldiers moving to the West, and on May 1st, 1945, he was captured at Lübeck by British soldiers.

Following the war,in the summer of 1946, Noll was enrolled in the Technical University of Berlin, with Mathematics as his main subject and Mechanics and Physics as secondary interests. In 1951, he obtained the degree of Diplomingenieur from the Department of Mathematics, equivalent at that time to the US degree of Master of Science in Applied Mathematics. During his undergraduate studies, Noll obtained a scholarship to study for one year at the University of Paris.

In 1952, Professor George Hamel, a leading German mechanician, was asked by Professor C A Truesdell to propose good young German mathematicians for doctoral studies at the Graduate Institute for Applied Mathematics at Indiana University in Bloomington. Walter Noll was one of those recommended and he duly arrived in the USA in September 1953. He asked Truesdell to be his PhD advisor and in August 1954, Noll successfully defended his thesis entitled "On the continuity of the solid and fluid states" before the appropriate examination committee.

There followed a brief sojourn at the University of Southern California in Los Angeles, before Noll accepted a position (in 1956) at what is now known as the Carnegie-Mellon University in Pittsburg. He was a Professor there from 1960 until his formal retirement in 1993.

The most influential period of Noll's career so far as rheology is concerned stretched from his PhD work to 1966, when the influential book *"Viscometric flows of non-Newtonian fluids"* (Coleman et al 1966) appeared. One of the main contributions to come from this period was the text *"The non-linear field theories of mechanics"* written jointly with his former teacher, C A Truesdell (Truesdell and Noll 1965). But the time is also remembered for the very productive collaboration with Bernard Coleman.

However, there were signs, as early as 1961, that the fruitful association with Coleman would have a limited life span and Noll's biographer, Ignatieff (1965, p127), gives as evidence the observation that at about this time the two scientists published separate papers on the same type of flow, which Coleman (1962a, b) called 'substantially stagnant motions', while Noll (1962) used the term 'motion with constant stretch history', the latter being the nomenclature favoured by subsequent workers in the field. After 1966, Noll and Coleman continued to consult each other on topics of common interest and, in 1990, Noll dedicated one of his more important later papers to Coleman on the occasion of his 60th birthday (Noll and Virga 1990). Furthermore, any suggestion of a rift or indeed of rivalry between the two has been vehemently denied (private communication between W Noll and KW).

In April 1955, Noll married Helga Schönberg in Berlin. They were both naturalized as American citizens in 1961. Helga died in tragic circumstances in 1976. In 1979, Noll married Mary (Resie) Strauss.

Bernard Coleman is younger than Walter Noll by five years. He was born in New York City in 1930. His undergraduate days were spent at Indiana University, where he graduated BS Cum Laude in 1951. He then moved to Yale University, where he obtained an MS in 1952 and a PhD in 1954, his research discipline at this time being Physical Chemistry. From 1954-1957, Coleman worked as a Research Chemist at the Carothers Research Laboratory of the DuPont Company. In 1957, he moved to the Mellon Institute in Pittsburg, initially as a Senior Fellow, before holding Professorships in Mathematics (from 1967) Biology (from 1974) and Chemistry (from 1984) at the Carnegie Mellon University. In 1988, he moved to Rutgers University in New Jersey as the J Willard Gibbs Professor of Thermomechanics. He also holds a Professorship in Mathematics at Rutgers.

Walter Noll's biographer concludes that Bernard Coleman's knowledge of mathematics was far from extensive until his move to Pittsburg, where he learned functional analysis through 'a great systematic effort' (Ignatieff, 1996, p217).

Unlike Noll, Coleman's interests in rheology continued after the ending of the partnership, and, although it is probably true to say that Coleman is still best known for his joint work with Noll, his collection of published works in Arch Rat Mech Anal **110**, 1990, 377-384 indicates an ongoing interest in various branches of rheology, and he was awarded the Bingham Medal of the Society of Rheology in 1984.

Bernard Coleman does not carry over his lucid writing style to his oral presentations at scientific meetings and he is often the bane of any strict time-keeping chairman.

4.11. Clifford A Truesdell

Clifford Ambrose Truesdell was born in Los Angeles, California, in 1919. He studied at the California Institute of Technology in Pasadena, where he received a BS degree in 1941 and an MS degree in 1942. He then moved to Princeton University as a graduate student and Instructor and obtained a PhD degree in mathematics in 1943. At Brown University in 1942, Truesdell came into contact with the plasticity-theory "establishment" led by William Prager; a mutual disregard seems to have ensued.

In the period between 1943-1950, there followed a series of fairly brief appointments, first as an Instructor at the University of Michigan, followed by a post at the Radiation Laboratory of MIT and then as chief of the theoretical mechanics subdivision at the Naval Ordnance Laboratory, before becoming head of the theoretical mechanics section at the Naval Research Laboratory.

In the 1946-1950 period, whilst still employed at naval establishments, Truesdell was also associated with the University of Maryland, rising to the rank of Associate Professor. In 1951, he became Professor of Mathematics at Indiana University, where he remained until his appointment as Professor of Rational Mechanics at the Johns Hopkins University in 1961. Truesdell retired in 1989 and continues to enjoy emeritus professor status at Johns Hopkins University.

Clifford Truesdell has received several honorary doctorates - Milan Polytechnic (1964), Tulane University (1976), Uppsala (1979) and Basel (1979). Other distinctions awarded include the Bingham medal in 1963, the Panetti International Medal and Prize (Academy of Science of Turin, 1967) and the Birkhoff Prize (American Mathematical Society and SIAM, 1978).

Truesdell founded the Society of Natural Philosophy and was chairman and then secretary of this organization for a total of about six years; the society seeks to promote rational mechanics. He is perhaps best known for his books and editorial work, especially with Springer-Verlag. He was also editor of some of Euler's collected works.

From 1952-1956, Truesdell was editor of the Journal of Rational Mechanics and Analysis; in 1957 he became editor of the rival Archives for Rational Mechanics and Analysis. He followed this with editorship of the new journal 'Archives for the History of the Exact Sciences' in 1960, despite his dislike of professional historians of science (Truesdell 1968, p2).

4.12. Arthur S Lodge

Arthur Scott Lodge was born in Liverpool, England in November 1922. Between 1941 and 1948, he was a student at Oxford University, receiving his BA in Mathematics and his DPhil in Theoretical Nuclear Physics. He joined the British Rayon Research Association in 1948 and worked initially with Karl Weissenberg. His early work in rheology involved the use of convected (or body) coordinates, very much along the lines proposed by Oldroyd in his 1950 paper (see, for example, Lodge 1951). He also began work on network models for polymeric liquids drawing on the ideas of Tobolsky. The research was eventually published in an influential paper in the Transactions of the Faraday Society (Lodge 1956) and the term 'Lodge rubberlike liquid' quickly became a part of standard rheological nomenclature.

In 1961, Lodge moved to the Department of Mathematics at the University of Manchester Institute of Science and Technology (UMIST) and the seven years he spent there were very productive. In particular, he and his research team carried out very detailed and innovative experiments on normal stress measurements in steady shear flow, and the research is still viewed as a paradigm of careful experimentation. The discovery of the 'hole-pressure error' was a noteworthy consequence of this research. Lodge later attempted to exploit the error to provide a measure of the elasticity in flowing liquids through what is often referred to as 'the Lodge Stressmeter'. In subsequent years, Lodge expended much effort in an attempt to commercialize the Stressmeter and to put the associated data interpretation on a sound footing; but the technique has not been sufficiently 'user friendly' to commend itself to a wide market.

Lodge's productive Manchester days also saw the publication in 1964 of his immensely influential book *"Elastic liquids"* (Lodge 1964), which was subsequently translated into Russian (1969) and Japanese (1975).

The 1965-66 academic year saw Lodge on sabbatical leave at the University of Wisconsin, and,in 1968, he decided to move permanently to the United States to become the world's first Professor of Rheology at the University of Wisconsin, Madison. He subsequently took out American citizenship.

Over many years, the Rheology Research Centre (RRC) at Madison, containing as it did some very distinguished rheologists, became an internationally renowned centre of excellence, which attracted many visitors and produced research of the highest quality. Lodge was the Chairman of the RRC from 1969 until his retirement in 1991.

In 1971, Arthur Lodge was awarded the Bingham Medal by the (US) Society of Rheology. He also received a Gold Medal from the British Society of Rheology in 1983, thus becoming the first and only rheologist to receive the highest award of the two oldest rheological societies. He was elected to membership of the National Academy of Engineering in 1992, an honour from his adopted country which gave him immense satisfaction.

Arthur Lodge is a devoted family man. He married Helen in July 1945 and they have two sons and one daughter. The Lodge household has long been viewed as a haven of culture, taste and old fashioned hospitality. At the height of his prime, Arthur Lodge was a fearsome debater at scientific meetings, who expected the same meticulous care from others as he demanded of himself. His penchant for limericks is legendary, e.g.

Weissenberg said "In liquids, confess
That large strains sometimes govern the stress."
Remarked Rivlin, "No drivelling!
In flows steady, though swivelling,
The elastic part must evanesce."

Then along came young Oldroyd, JG.
"Things aren't quite as they seem, now", said he;
"With components convected
(As H Hencky selected),
One obtains a new D by Dt."

(From "Karl Weissenberg Commemorative Scientific Essays" 1973, Ed J Harris.)

4.13. David V Boger

David Vernon Boger was born in Kutztown, Pennsylvania, on 13th November 1939. He obtained his BS in Chemical Engineering in 1961 from Bucknell University, Pa, and then his MS (1964) and PhD (1965) from the University of Illinois in Urbana.

Boger's first University appointment was as Lecturer in Chemical Engineering at Monash University in Victoria, Australia (1965). The secretary to the then Department Chairman was Liz Mannix, who was to become Mrs Boger in October 1967.

Boger had a successful research career at Monash; he was promoted to Senior Lecturer in 1971 and to Reader in 1980. In 1982, he moved to the nearby University of Melbourne as Professor of Chemical Engineering, where his research continued to flourish.

David Boger has given his name to an important test fluid, the so-called Boger fluid, 'a highly-elastic constant-viscosity fluid' first introduced in the late seventies (Boger 1977/78, cf Walters 1979.)

Boger and his Monash research team had conducted many experiments on shear-thinning polymer solutions and had realized that interpretation of the growing mass of experimental data was made difficult by the dual complications of fluid inertia and shear thinning. The former could be easily overcome by increasing the viscosity levels, but the shear thinning influence was more difficult to accommodate. However, some years earlier, the Monash team had flirted with the drag reduction phenomenon and had realized, with others of course, that small concentration of polymer can sometimes have a significant influence on flow characteristics, with very little change in viscosity levels.

This experience led the team, more particularly David Boger and Rod Binnington, to use the most viscous Newtonian liquid they could find (corn syrup) and to attempt to get very small concentrations of their favourite polymer, polyacrylamide, into solution. This was accomplished by first dissolving the polyacrylamide in water and then adding the water to a very viscous corn syrup. The experiment was immediately successful: "What we had was a material whose viscosity was constant and where we were generating normal stresses, which were as high as ten times the shear stress. There it was: a constant viscosity elastic liquid". Since that time Boger fluids have been popular test fluids in many esoteric rheological studies.

At the University of Melbourne, Boger's research interests broadened considerably to encompass coal suspensions, waxy crude oils, ceramics, hydrocolloids and all manner of particulate fluids. He has received many national and international honours for his research endeavours; in 1989 he became a Fellow of the Australian Academy of Technological Sciences and Engineering and in 1993 he was made a Fellow of the Australian Academy of Science. In 1994 he received the Australian Society of Rheology Medallion and, in 1995, he was awarded the prestigious Walter Ahlström Environmental Prize by the Finnish Academies of Technology for "significant technological achievements in the areas of industrially applicable advances in the use of energy and raw materials, and for the minimizing of environmental impact".

David Boger is often called on to give plenary lectures at international conferences; he is known as an enthusiastic lecturer with a colourful turn of phrase, which is more characteristic of his adopted Australia than the land of his birth. He can be combative in discussions following lecture presentations, but he is always scrupulously fair and objective. He has a disarming sense of humour and a great love for the outdoors.

David Boger became an Australian citizen in July 1997.

4.14. BKZ

Barry Bernstein, Elliot Kearsley and Louis Zapas at the Rochester Society of Rheology meeting in 1991.

Professor Barry Bernstein, Dr Elliot Armstrong Kearsley and Dr Louis J Zapas were influential US rheologists in the post Second World War period. They came together at the National Bureau of Standards (NBS; now the NISR) in Washington DC and collaborated on a number of rheological topics. They are best known for their work on deriving an integral constitutive equation for polymer liquids, which has become widely known as the BKZ model.

Barry Bernstein, the main motivator of the collaborative venture, was born in Brooklyn, New York in 1930. He received a BS in Mathematics from the City College of New York in 1951, before moving to Indiana University, where he obtained an MA in Mathematics in 1954 and a PhD in Mathematics/Mechanics in 1956. Bernstein was associated with the Naval Research Laboratory from 1951-61 and with the NBS from 1961-65. He then moved to Purdue University, before taking up a tenured Professorship at the Illinois Institute of Technology in 1966.

Elliot Kearsley was born in Springfield, Massachusetts in 1927. He served in the US Army during the Second World War. He then subsequently studied Physics at Harvard University, earning an AB (magna cum laude) in 1949 and an MA in 1951. In 1955, he was awarded a PhD in Physics by Brown University.

Kearsley was employed at the Bendix Aviation Research Laboratory in Detroit from

1953 to 1955. He then served as a member of the technical staff of the Committee on Undersea Warfare of the National Research Council - National Academy of Sciences. In 1957, Kearsley joined the NBS as a Project Leader in the Rheology Section of the Heat and Power Division. He became Chief of the Mechanical Properties Section of the Polymer Division in 1973. He left that position in 1976 to serve as Liaison Scientist for the Office of Naval Research in Tokyo Japan. He returned to the NBS in 1978 and remained there until his formal retirement in 1981.

Louis J Zapas was the most flamboyant of the NBS triumvirate. After receiving an MS in Chemical Engineering from the University of Pittsburgh in 1949, he worked at the Mellon Institute. From 1957 to 1961 he was at the Washington Research Center of W R Grace and Co. In 1961, he joined the NBS and remained there until his retirement in 1991, the year he was awarded the Bingham medal by the Society of Rheology. The citation referred to his "extraordinary contributions to the field of rheology, especially in the development and experimental testing of the BKZ constitutive equation".

4.15. Doi and Edwards

1995. Sir Sam Edwards and Masao Doi relaxing after a dinner at Gonville and Caius College, Cambridge.

The association of Sir Sam Edwards and Masao Doi in the mid-seventies had an enormous impact on polymer melt rheology and its influence is still apparent today. At that time, Edwards was already a senior scientist on the international stage, whereas Doi was a junior staff member at a Japanese University; but the association was a true meeting of minds. It resulted in the publication of a series of seminal papers on the 'reptation model' for polymer melt flow, followed later by an influential book entitled *"Theory of polymer dynamics"*. The joint work of Doi and Edwards during this period was a catalyst for a surge of interest in the reptation model for polymer melts and it catapulted Doi into international prominence.

Sam Edwards was born in Swansea, Wales, on February 1st, 1928. He attended the local grammar school, before going up to Gonville and Caius College, Cambridge, to pursue undergraduate and postgraduate studies. His PhD thesis was finally completed at Harvard University, during a stay in the US which also took in the Institute of Advanced Studies at Princeton.

Edwards returned to the UK in 1953 to a post at the University of Birmingham. In 1958, he moved to the University of Manchester, and from 1963 to 1972 he held the post of Professor of Theoretical Physics. He was elected to a Fellowship of the Royal Society in 1966.

Edwards renewed his association with the University of Cambridge in 1972 on his appointment as the John Humphrey Plummer Professor. He became the Cavendish Professor of Physics and Director of the Cavendish Laboratory in 1984, following in the footsteps of such renowned scientists as J C Maxwell, Lord Rayleigh, J J Thomson and Lord Rutherford. He was knighted in 1975.

Sir Sam has been one of the most distinguished and energetic British scientists of the 20th century. He has served on numerous national and international scientific committees; he has received several honorary doctorates; he has been awarded many scientific honours, including a gold medal from the British Society of Rheology.

Since his formal retirement from the Cavendish Laboratory, Sir Sam has continued to conduct research of distinction. His penetrating physical insight and impressive mathematical skills are still being brought to bear on a wide range of scientific problems relating to material behaviour. He is still in demand as a plenary speaker at international conferences.

Sir Sam's other attributes are ably summed up by a quote from a speech made on the occasion of his admission to an Honorary Fellowship of the University College at his home town of Swansea: "The panoply of honours, degrees and awards could create in one's mind the impression that Sir Sam might well conform to the popular image of the great scientist: remote, ascetic and icily detached from the real world. Nothing could be further from the truth. He is eminently approachable, friendly and ebullient, with a connoisseur's appreciation of fine wines, food and art. He is a first rate raconteur, gifted with a subtle pawky sense of humour, and a deep concern for humanity".

Masao Doi was born twenty years later than Sir Sam Edwards, on March 28th, 1948, at Aichi-prefecture, Japan. He was educated at the University of Tokyo, obtaining an MSc degree in Applied Physics in 1973 and a Doctor of Engineering in 1976 for a thesis entitled *"Statistical theory of the molecular entanglement in polymer systems"*.

In 1976, Doi obtained a Fellowship from the UK Science Research Council to work first at Imperial College and then at the University of Cambridge. The initial contact with Sam Edwards was made while Doi was at Imperial College and this flourished following the move to Cambridge in 1977. (Doi was to spend another year at Cambridge in 1984-85 as a Fellow of the Japan Society for the Promotion of Science.)

Doi returned to Japan in 1978 as an Associate Professor in Physics at Tokyo Metropolitan University. He stayed there for nine years, before moving to Nagoya University in 1989 as Professor of Applied Physics.

Doi soon became a scientist of international repute in his own right and he remains an influential and respected worker in polymer melt rheology and related fields. He is a lucid and articulate lecturer, who is in constant demand as a plenary speaker at scientific meetings; he gave one of the invited opening lectures at the Quebec International Congress in 1996. Doi has a steely respect for scientific integrity, which is often masked by an infectious personality and disarming smile.

4.16. Giuseppe Marrucci

Giuseppe (Pino) Marrucci was born in Italy on April 16th, 1937. His academic career has been very much associated with the University of Naples. It was there that he received a doctorate in Chemical Engineering in 1961. There followed a period as Assistant to the Chair of Industrial Chemistry, before he became Associate Professor of Chemical Engineering in 1965. Marrucci later moved to the University of Palermo in Sicily to take up a Chair of Principles of Chemical Engineering. Italian bureaucracy ensured that the move to Palermo was not smooth and the 1970-71 academic year saw Marrucci virtually holding down two faculty positions, one at Naples and one at Palermo, something that required serious commuting.

Marrucci remained at Palermo until 1976. He then returned to the University of Naples to hold chairs in non-Newtonian Fluid Mechanics and Thermodynamics. He has remained there ever since.

Marrucci's early association with the late Gianni Astarita was influential and helped to establish the University of Naples as a centre of excellence in rheology, something it has maintained in subsequent years. Those early years saw Marrucci's research concentrated on phenomenological approaches to non-Newtonian fluid mechanics and this culminated in the publication of a research text, written jointly with Astarita (Astarita and Marrucci 1974).

Marrucci's move to Palermo involved more research in *continuum* aspects of non-Newtonian flow, but, importantly, a new research interest was developed, involving the *molecular* modelling of the rheology of polymeric liquids, something that has become a dominant feature of Marrucci's research in later years. His present activity is concentrated on the molecular modelling of the rheology and rheo-optics of polymers, either ordinary or liquid crystalline, and polymers with localized interactions.

During his academic career, Pino Marrucci has travelled widely, with numerous visiting positions at foreign universities, including Delft, Delaware and Berkeley. He is a lucid lecturer and expert communicator. He remains a person of warmth, charm and culture.

Chapter 5

From Continuum Theory to Microstructure (and Vice Versa)

5.1. Developments

As we indicated in Chapter 1, the early attempts by Navier (1823) and Cauchy (1827) to discover the gross behaviour of matter, by beginning from a postulated microstructure and working out the consequences, were reasonably successful, but the method fell out of favour when experiments showed that Poisson's ratio did not necessarily have the value of 0.25. For gases, the work of Maxwell (1867) and Boltzmann (1896) showed that a great deal of understanding could be reached from microstructural mechanical axioms. Nevertheless, the difficulties that Boltzmann had in getting molecular ideas accepted around 1900 have to be noted. It was only the work on Brownian motion (named after Robert Brown, a Scottish botanist, who noticed the violent, irregular motion of particles in water in 1848) that eventually led Perrin to show that no other reasonable hypothesis than the molecular one could explain the phenomena observed by Brown and others. Perrin was awarded the Nobel prize for this work in 1908 and one can conclude that, with minor retrogressions, molecular reality was fully accepted by this date.

From the point of view of rheology, we can date the beginning of currently useful microstructural theories to the work of Albert Einstein (1879-1955). In 1905, Einstein produced a theory of Brownian motion and diffusion. He showed how this theory could be used to determine Avogadro's number and also to predict the viscosity of a dilute suspension of spheres. If the volume fraction of the spheres is c, Einstein (1905) showed that the effective viscosity was raised from the solvent viscosity η_0 to η, where

$$\eta = \eta_0(1 + 2.5c). \tag{5.1}$$

This formula is still of great utility for very dilute suspensions; for larger values of c the computations are much more complex, as we shall discuss below. Further results on the general theory of random fluctuations at equilibrium were published in 1906, including rotational diffusion.

In the same year (1906), the Polish physicist, Marian von Smoluchowski (1872-1972), who was born and educated in Austria, gave a collisional theory of Brownian motion, considering acceleration and other changes as being due to the collision of smaller impacting particles. Paul Langevin (1872-1940) formed a stochastic differential equation for a Brownian particle, so that he wrote, for the particle under consideration (Langevin 1908):

$$m\frac{d\boldsymbol{v}}{dt} = -\boldsymbol{f} + \boldsymbol{\Delta}, \tag{5.2}$$

where $md\boldsymbol{v}/dt$ is the mass-acceleration of the particle, $\boldsymbol{f}$ is a retarding force (due to, for example, fluid viscosity) and $\boldsymbol{\Delta}$ is a stochastic force, essentially a representation of the other molecules bombarding the chosen particle. He was able to show that the particle diffusion obeyed the same laws as had been described by Einstein in 1905 (Langevin 1908). Von Smoluchowski in 1915 then managed to write down the appropriate diffusion equation in the presence of external forces; he also discussed solutions for various boundary conditions.

In 1917, Max Planck (1858-1947) derived the so-called Fokker-Planck equation, which described the probability density of a system configuration; he was motivated by Adriaan Fokker's earlier work on electric dipoles. The Dutch physicist, L S Ornstein (1880-1941), clarified further the Fokker-Planck work in 1919; it was left to the Swede, Oskar Klein (1894-1977) to set out the Fokker-Planck equation for a system of interacting Brownian particles subject to external forces, thus setting the stage for rheological applications (Klein 1921).

The above essential theoretical groundwork on probability was extremely well expressed in 1943 in a famous paper by Subrahmanyan Chandrasekhar[1] published in 'Reviews of Modern Physics' (1943).

5.2. Macromolecular Hypothesis

It has to be emphasised that the widespread recognition of molecules only dates from about 1908, and the concept of very large molecules, or macromolecules, followed about 20 years later; the principal actor in this discovery was Staudinger.[2]

During his sojourn in Karlsruhe, Staudinger synthesized isoprene by a new path, and used polyisoprene as a test of whether the "aggregate" or the "long-chain-molecule" theory of polymer structure was correct; in 1920 he championed the latter. Six years later, he

[1]S Chandrasekhar (1910-1995), who was popularly known as Chandra, was interested in astronomical applications. Nevertheless, his clear exposition has assisted workers in other fields, especially in electrical circuit theory and rheology, to get started in the subject of stochastic mechanics. Clear exposition of many fields was a hallmark of Chandra's work; his book on Hydrodynamic Stability (1961) is also of great interest to rheologists. A somewhat retiring figure, Chandra worked at Cambridge University in England, where he had severe differences of opinion with James Jeans. He then spent the rest of his career based at the Yerkes Observatory of the University of Chicago. His last book, published in 1995, entitled "The Principia for the Common Reader" is an exposition of Isaac Newton's work on mechanics with proofs which are understandable to the modern reader. He was awarded the Nobel Prize in Physics for his work showing that stable stars have an upper limit of mass of about 1.4 sun masses. A sympathetic biography is that of Wali (1991).

[2]Hermann Staudinger was born in Worms, Germany, on the 23rd March 1881. He wished to study botany, but was persuaded by his father to study organic chemistry first. He studied at Munich and Halle, obtaining his doctorate in 1903. After a spell in Strasbourg, where he made the discovery of ketenes (coffee aroma, 1907), he went to Karlsruhe, then to the ETH in Zürich, and finally to Freiburg-im-Breisgau in 1926, where he died in 1965.

predicted that long molecules would be important in biology. Staudinger and his co-worker used the words "macro-molecular association" for the first time in 1922, and the word "makromolekül" in 1924. Vigorous opposition to these ideas soon followed. Needing to pursue his studies, and having no money for a centrifuge, Staudinger used viscometry for the study of specific viscosity and molecular weight. In due course, opposition to macromolecules died away and Staudinger was awarded the Nobel Prize for Chemistry in 1953, when he was seventy-two. His view that giant molecules existed (as opposed to mere aggregates of small molecules) had, as we have already stated, run into very severe opposition. In 1932, his book (Staudinger 1932) appeared, but, even as late as 1937, Houwink (1937 p 212) had clear reservations about the work of Staudinger, as well as that of Mark and Kuhn.

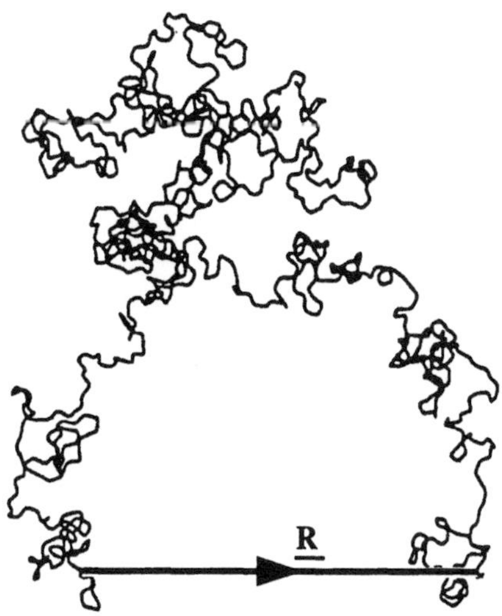

Fig 5.1. Form of one configuration of a random-walk 1000-link chain molecule according to the statistical theory. The end-end distance (vector ***R***) is about 10% of the root-mean-square dimension of the molecule (from Treloar 1958).

Once the idea of giant molecules had been accepted, their configuration needed to be considered. **Werner Kuhn** began to consider this problem in 1932-3 and initially supposed that the molecules were of a dumbbell or shishkebab form, with a rigid backbone. When his predictions did not agree with viscosity and streaming birefringence measurements, Kuhn then produced his 1934 paper, showing that molecules had a random, coiling, thread-like form, governed by Boltzmann's entropy/configuration law, $S = k \log W$.[3] Figure 5.1 shows the form of such a molecular model. There were others who produced vital information on these questions. Meyer, von Susich and Valko (1932) became convinced that rubber elasticity was due to atomic motion in long molecules, and LRG Treloar (1943)

[3]This formula relates the entropy (S) to the number of possible states (W) of a system; the Boltzmann constant k has the value $1.3807 \times 10^{-23} J/K$.

(see also Scott Blair 1949, p39) believed that these authors were the first to state the idea. Wöhlisch (1939) claimed to have put forward this idea in 1926, but K H Meyer (1939) essentially dismissed these claims; Meyer stated that Wöhlisch's theory was wrong, and it was. Meyer et al (1932) seemed to have considered lateral pressure between molecules, rather than direct tension along a single molecule, to be the result of molecular motion. In 1934, Eugen Guth and Hermann Mark produced a formula for tension in a molecule based on entropy change.

In 1936, Werner Kuhn put all these ideas on a quantitative basis. Many of our current terms (such as"Kuhn step"and "Kuhn length") recognize Kuhn's contributions . He also realized that coil expansion would arise from excluded volume effects. Unfortunately, his result that intrinsic viscosity was proportional to molecular weight raised to the power 0.84 did not agree with the then accepted linear law of Staudinger (1933).

Kuhn, Guth, Mark, and Meyer were all aware of one another's work, and so it is difficult to assign a strict priority to any group, since each contributed portions to the overall solution, with Kuhn eventually synthesizing the work in 1936.

Meyer and Ferri (1935) discussed the kinetic theory of elasticity; they stated that the principal valencies were oriented by pulling and that the transverse motion of the chains caused a pressure normal to the chains. They also say that F Busse (1932) had "a peu près simultanément une théorie analogue". In the longer run, this idea of normal pressure was superseded by Kuhn's concepts. Further contributions were made by Treloar (1943) who showed that, provided some approximations made by Kuhn were replaced by exact expressions, the theories of Kuhn (1936) and Wall (1943) agreed. John G Kirkwood[4] (1945) was able to say at this time "The statistical theory of the equilibrium elastic behaviour of rubber has contributed greatly to our understanding of the structure of cross-linked polymers, although, from the quantitative point of view, it leaves something to be desired. The basic assumption of the theory is that the resistance offered to deformation by a cross-linked polymer network is principally due to the decrease in configurational entropy attending the elongation of the chain segments of the net". Detailed reading of this paper shows in fact that there were still many unresolved questions that would remain for another decade or so; see also Kuhn and Grün (1946), James and Guth (1943) and especially Morawetz (1995) for a more complete discussion.

5.3. Dilute-Solution Theories

At this point, it is convenient to sum up the contributions of Werner Kuhn: He developed the random-walk model for polymers in solution; his derivation of the end-to-end distance vector in terms of bond lengths and angles is now classical. Kuhn also introduced the relation between the force at the chain ends and the distance between them,

[4]John Gamble Kirkwood was born in 1907. He received an SB degree from the University of Chicago in 1926 and a PhD degree from M.I.T. in 1929. Later, from 1931-1932, he worked as an International Research Fellow with P Debye in Leipzig and in the following year as a Research Associate at M.I.T. His first Faculty position was as Assistant Professor of Chemistry at Cornell. There followed a brief sojourn at the University of Chicago, before he returned to Cornell in 1938 as the Todd Professor of Chemistry. He stayed there until 1947, when he moved to Cal. Tech. as the A.A. Noyes Professor. In 1951, he moved again, this time to Yale University, where he remained until his death in 1959. From 1955 to 1958 he was Foreign Secretary of the National Academy of Sciences (see Bird 1988).

thus introducing the linear spring concept for a macromolecule in the form:

$$\boldsymbol{F} = -3NkT\boldsymbol{R}/(Na)^2, \tag{5.3}$$

where **F** is the end-end force, N is the number of links, length a, in the chain, k is the Boltzmann constant, T the absolute temperature and **R** is the end-end vector of the molecule. The law (5.3) has been used in many subsequent papers which model polymer flow, and it is often referred to as the Gaussian chain model. Kuhn also recognized the importance of the excluded-volume effect in polymers, a subject tackled much later by P J Flory (see, for example, Flory 1969).

In 1944, having been prevented from working on nuclear physics in wartime, Hendrik Antony Kramers (1944) developed a kinetic theory for dilute polymer solutions undergoing potential flow. Kramers obtained the zero-shear viscosity and considered some very general polymer models, including freely jointed bead-rod chains, ring molecules and branched molecules. This paper was a key contribution to the micromodelling of polymers. In 1948, Kirkwood began his extensive work on the subject of dilute solution theory. He and Jacob Riseman (1948) introduced hydrodynamic interaction between polymer and solvent. Following this work, P E Rouse (1953) studied a chain molecule consisting of N Gaussian springs and $N+1$ beads. He was able to use matrix methods to decouple the modes and describe the entire frequency response in small-strain oscillatory flow. These results are of great interest, and subsequently found application to *concentrated* solutions. Bruno Zimm (1956) introduced the idea of averaged hydrodynamic interaction into the model; this is another idea of long-term interest.

With these results and other relevant work (Wall 1943, Eyring 1932), one can see that the problems of small-strain motion, described by the linear storage and loss moduli, were essentially resolved for very dilute solutions. (For less dilute solutions, it is doubtful if the time scales of the Rouse model are applicable.) In a sense, the completion of this part of the work achieved the goal of John D Ferry (1970) whose aim was to use "mechanical spectroscopy" to probe the molecular state. Ferry's book has been through many editions and is still a key reference. Additionally, mention should be made of the extraordinary work of P J Flory[5] (1953, 1969).

Concerning progress in the case of more general flows, it is necessary to refer to a variety of models, beginning with the rigid dumbbells of Werner Kuhn (1932) and graduating to the flexible dumbbells of J J Hermans (1943) and the chain models of Kramers (1944). There have been many elaborations, which are fully detailed in the book of Bird et al (1987b).

When dealing with micromechanics, it is generally necessary to:

1. Compute the probability function ψ for a given configuration.
2. Given ψ, compute the average stresses in the fluid.
3. If possible, find a constitutive equation.

[5]Paul J Flory was born on the 19th June 1910 in Sterling, Illinois. He completed his PhD at Ohio State University in 1934 and then joined the DuPont Company, working with W Carothers, the inventor of nylon and neoprene. During the Second World War, Flory worked in industry, before moving to Cornell University in 1948 and then to Stanford in 1961. He was awarded the Nobel prize for Chemistry in 1974. Flory died in 1985.

The computation of the distribution function ψ, once the Fokker-Planck equation is set up for the system, is not a trivial matter.

The computation of the stresses goes back to Cauchy (see Love 1952), who considered the interparticle forces crossing a plane, when producing his theory of molecular elasticity. This argument considers the number of forces crossing an area, and produces the following result for the molecular stress contribution $\boldsymbol{T}^{(m)}$:

$$\boldsymbol{T}^{(m)} = n_0 \int \psi dV \boldsymbol{F}\boldsymbol{R}, \tag{5.4}$$

where $\mathbf{F}$ is the force in the connector cutting the plane, $\mathbf{R}$ is the length of the connector, and dV is the element of configuration space. n_0 is the number of, say, dumbbells per unit volume. A further isotropic contribution $(-n_0 kT\boldsymbol{I})$ due to momentum flux is applicable for flowing fluids (Bird et al 1987b).

Similar results were used by Kramers (1944) and have been elaborated by others. Rivlin (1949b) made an early attempt to find normal stresses in shear flow from a macromolecular solution theory, but the results did not accord with experiments in general form; he found $N_1 = 0,\ N_2 \neq 0$.

Another approach was taken by **Hanswalter Giesekus** (1966), who showed that computation of the average ($< \boldsymbol{RR} >$) for flowing linear elastic dumbbells gave the following extraordinary formula:

$$\frac{\Delta}{\Delta t}(< R_i R_j >) + \frac{12kT}{\zeta N a^2} < R_i R_j > - \frac{4kT}{\zeta}\delta_{ij} = 0, \tag{5.5}$$

where the end-to-end vector of the dumbbell is $\mathbf{R}$ (components R_i), ζ is the bead friction factor, Na is the total length of the spring, a is the length of a submolecule, and kT is the Boltzmann-temperature factor. The $\Delta/\Delta t$ operator is the "upper-convected" time derivative.

Use of (5.4) and (5.5) permits us, in some cases, to derive a constitutive equation for the stresses. If we rewrite (5.4) to take account of the isotropic flux term, we obtain, for a linear Hookean spring:

$$\boldsymbol{T} = \boldsymbol{T}^{(m)} + n_0 kT\boldsymbol{I} = 3n_0 \frac{NkT}{(Na)^2} < \boldsymbol{RR} > . \tag{5.6}$$

Thus the terms $< \boldsymbol{RR} >$ can be eliminated from (5.5) in terms of $\boldsymbol{T}$, to give a constitutive equation of the form:

$$\lambda \frac{\Delta \boldsymbol{T}_{ij}}{\Delta t} + \boldsymbol{T}_{ij} = n_0 kT \delta_{ij}, \tag{5.7}$$

or, in terms of $\boldsymbol{T}^{(m)}$,

$$\lambda \frac{\Delta \boldsymbol{T}^{(m)}}{\Delta t} + \boldsymbol{T}^{(m)} = 2\eta^{(m)}\boldsymbol{d}, \tag{5.8}$$

where the molecular viscosity $\eta^{(m)} = \lambda n_0 kT$. This equation is identical in form to the Upper-Convected Maxwell (UCM) model mentioned in Chapter 4 (eq 4.13 with $\lambda_2 = 0$). The total stress $\boldsymbol{\sigma}$ consists of

$$\boldsymbol{\sigma} = -p\boldsymbol{I} + 2\eta_s \boldsymbol{d} + \boldsymbol{T}^{(m)}, \tag{5.9}$$

where η_s is the solvent viscosity and p is a pressure.

In our opinion, this result is one of the most exciting ever to appear from micromodelling. In turn, it was generalized by Lodge and Wu (1971) to a Rouse chain model, resulting in a total molecular stress made up of modal contributions of the form (5.8). These models suffer from the known defect of the UCM - that the elongational viscosity diverges at an elongation rate of $1/2\lambda$. This is because (in the linear Hookean dumbbell model), hydrodynamic forces are eventually able to completely overwhelm the spring forces for high extension rates, and the bead separations become very large. To prevent this, one can use a spring of limited length, reflecting the case of real molecules. In such a molecule, we expect, following Flory (1969), the spring law

$$\boldsymbol{F} = \frac{kT}{a}\left(\frac{\boldsymbol{R}}{R}\right)\mathcal{L}^{-1}(R/Na), \tag{5.10}$$

where $\mathcal{L}^{-1}(x) = \coth x - \frac{1}{x}$.

The $\mathcal{L}^{-1}$ function is the inverse Langevin function, named after Langevin (1908), who used it in the study of magnetic domains. Instead of the inverse Langevin function, several springs of limited length have been proposed; the most popular variation is that of Harold Warner (1972), who used

$$\mathrm{F} = \frac{3kT\boldsymbol{R}}{Na^2}/\left(1-\left(\frac{\boldsymbol{R}}{Na}\right)^2\right). \tag{5.11}$$

Clearly there are detailed differences in behaviour between equation (5.10) and (5.11). Because $\boldsymbol{F}$ is no longer proportional to $\boldsymbol{R}$, it is not possible this time to determine a constitutive equation. R I Tanner (1975) observed that, if one assumes that the distribution function ψ is localized, and can be represented as a delta function at some unknown location, one could proceed to an approximate constitutive equation, which will be exact for both very weak and very strong[6] steady flows. (It is also possible to consider the distributions as a narrow set of Gaussians, with the same results.) The constitutive equation is of the form

$$\lambda\frac{\Delta}{\Delta t}(\boldsymbol{T}/K) + \boldsymbol{T} = \eta_0 kT\boldsymbol{I}, \tag{5.12}$$

where, for the Warner spring, the function K is

$$K = 1 + (tr\boldsymbol{T})/3\eta_0 NkT. \tag{5.13}$$

See also Armstrong and Ishikawa (1980) and Chilcott and Rallison (1988) for further development of this area.

5.4. Concentrated Solutions and Melts - The Network Theory

The preceding text shows that, by the early 1940's, the entropic spring model of polymer molecules was firmly established and good progress had also been made towards a theory of (solid) rubber elasticity. The article by James (1947) is a clear expression of one

[6]'Strong' in this connection means great disturbance of the microstructure.

form of the "Gaussian network" theory. From the rheological point of view, little had been done for fluids, except for dilute solution theory. In 1946, M S Green and A V Tobolsky set out to consider the relaxation processes in "networked" polymers. At the outset, they considered "permanent" bodies, whose construction did not change with time; they said "... the internal processes of relaxation that are occurring in the material are characterised by equal rates of breaking and reforming of the bonds which constitute the network chains...". They assumed a single relaxation time, regarding the extension to many times as being obvious. They gave reasons for these assumptions, citing experiments on polysulfide rubbers. Using the kinetic theory of rubber elasticity, they assumed the stress σ was given by

$$\sigma/ckT \begin{array}{l} = \sum n_i(\ell^2/\ell_i^2 - \ell_i/\ell) \text{ (tension)}, \\ = \sum n_i(\gamma - \gamma_i) \text{ (shear)}, \end{array} \tag{5.14}$$

where c is a constant, possibly, but not necessarily, equal to unity. n_i denoted the number of chains that were formed when the sample had a length ℓ_i or shear γ_i. Making the assumption that the number of bonds (n) obeys the kinetics

$$\frac{dn}{dt} = -k'n, \tag{5.15}$$

where k' is a constant, they were able to show that, for the shear case, for example,

$$\sigma/ckT = n_0 \exp(-k't)\,[\gamma - \gamma_{eo}] + \int_0^t n_0 k' \exp(-k'(t-t'))] \times [\gamma(t) - \gamma(t')]dt'. \tag{5.16}$$

(Here γ_{eo} is the shear strain at $t = 0$). There is a corresponding equation for extension. Thus, an integral form of constitutive relation was deduced. Not content with this result, the authors differentiated (5.16) to produce results like (for shear)

$$\frac{d\sigma}{dt} + k'\sigma = G\frac{d\gamma}{dt}, \tag{5.17}$$

which is a Maxwell equation; for extension, they obtained a different (nonlinear) form. By a cumbersome set of energy-based arguments, they produced the result (G/k') for the shear viscosity, and the result

$$\sigma = 3\left(\frac{G}{k'}\right)\left(\frac{\dot{\ell}}{\ell}\right) \tag{5.18}$$

for extensional flow where $\dot{\ell}$ is the rate of extension.

The Green and Tobolsky paper therefore contains a new departure for rheology, in that macroscopic behaviour for non-cross-linked, non-dilute media was deduced, beginning from rubber-like theory. They said, in their Appendix B, that "it will be very easy to extend the ideas developed in the main part of this paper to the most general three-dimensional strain history". Green and Tobolsky (1946) made a start at such a theory, and showed that the extra-stress tensor must contain time derivatives of the convected type. A brief mention should be made of the extraordinary paper of Loring (1950), entitled *"Theory of the mechanical properties of hot plastics"*. This paper used rubber-like theory with large strain and considered a theory close to that developed by Lodge (1956), obtaining a Maxwell-like behaviour in shear. He also suggested using elastic "potentials" of the general power-law form (i.e. proportional to the (principal stretches)^n) in a manner used later by Ogden (1984) for rubbers.

The theory then lay dormant until Lodge (1954, 1956) and M Yamamato[7] (1956, 1957, 1958) took up the mathematical formulation of what has since become known as network theory. An earlier paper by Yamamoto (Yamamoto and Inagaki 1952) is written in Japanese and is not widely known. Two important papers by Yamamoto and Lodge, both published in 1956, yielded identical results, under the Green-Tobolsky assumptions about network strand breaking and reformation. However, Yamamato went further and introduced more general kinetics for breaking and reformation of network "junctions". Following these works, it was shown that the Green-Tobolsky model could be precisely formulated as a constitutive equation connecting the (extra) stress $\boldsymbol{T}(t)$, and the history of the Finger strain tensor $\boldsymbol{C}^{-1}(t')$ (measured relative to the current configuration at time t) in the familiar form:

$$\boldsymbol{T} = \int_{-\infty}^{t} m(t-t')\boldsymbol{C}^{-1}(t')dt'. \tag{5.19}$$

In the initial concept, $m(t-t')$ was a single exponential $m(t)$, given by a constant $\times \exp(-k't)$, but 5.19 is a generalization introduced by Lodge (1956) to encompass a more general (linear) relaxation behaviour. The importance of the difference between shear and elongational behaviour was, following Green and Tobolsky, brought firmly to the attention of rheologists by Lodge; his book (Lodge 1964) has been exceptionally influential in this respect. From these pioneering works, a number of constitutive relations have been developed; for example, for network rupture (Tanner 1969) and non-affine motion (Johnson and Segalman 1977, Acierno et al 1976, Phan-Thien and Tanner 1977, Phan-Thien 1978). An interesting set of models was given by Kaye (1966), who assumed that network rupture depended on local stress. Takano (1974) also investigated various forms of Yamamoto's ideas. An account of these theories is given by Bird et al (1987b), Wiegel (1969), Wiegel and de Bats (1969) and by Tanner (1988). Mention should also be made of the extensive work of M H Wagner (from 1978 onwards), who used these theories to describe experiments, and of the book on constitutive equations of Larson (1988).

5.5. Reptative Rheology

Although the network model of polymers described above was, and still is, very useful, there were problems in defining the nature of the strand interactions ("entanglements"), and a search for other models began in the 1970s. The most successful ideas were "reptative", based on tube models.

[7]Misazo Yamamoto was born in March 1928, the son of a Buddhist priest. His first name, which means 333 in Japanese, was given to him by his parents since he was born on the third day of the third month of the third year of the reign of Emperor Hirohito. He studied Physics at Kyoto University, where he received BS, MS and PhD degrees. He later became an Associate Professor in the kosa of Professor S Oka in the Department of Applied Physics of Tokyo Metropolitan University. Yamamoto was lame from childhood, having suffered from infantile paralysis. He was nevertheless always filled with good will and humour. He died of pneumonia on May 14th, 1974. The Japanese Society of Rheology published a memorial issue for him of Nihon Reoroji Gakkaishi (Volume 2, Issue 3, 1974).

The "tube" model was originally proposed for rubbers (Edwards 1967). This model of polymers assumes that, due to topological constraints, chain motion is essentially confined to a tube-like region; the "tube" is made up of the surrounding chains (Doi and Edwards 1978, 1986). This model has been very successful in explaining many properties of polymers and has generated an enormous literature; some properties (e.g. steady shear viscosity) are not yet satisfactorily explained. The experimental work of Osaki[8] and coworkers has been important in estimating the utility of the reptation models (cf. Doi and Edwards 1986).

The tube model, invented for rubber theory, was also applied by de Gennes[9] (1971) to uncrosslinked systems such as polymer melts. Here, one needs to recognize that the "tube" itself changes with time. De Gennes (1979) imagined a chain under the influence of Brownian motion moving through a fixed network. The chain "crawls" through the tube space and creates a new "tube" as it goes; de Gennes called this type of motion "reptation", after the Latin "reptare" (to creep).

Chapter 7 of Doi and Edwards (1986) is devoted to polymer liquid viscoelasticity and the following successes are noted:

1. Klein and Briscoe (1978, 1979) measured the diffusion constant (D_G) of deuterated polyethylene molecules in a polyethylene matrix and found $D_G \alpha M^{-2}$,where M is the molecular weight. This result agrees with the prediction of reptation theory.

2. The observed stress-optical law is predicted by the theory.

3. A reasonable agreement with the linear viscoelastic properties of polymers is found.

There are however, some areas of disagreement; we have already mentioned the viscosity function, $\eta(\dot{\gamma})$, which is as yet not satisfactorily predicted; (the stress drops away much too fast as $\dot{\gamma}$ increases); the viscosity-molecular weight dependence is predicted to be $\eta \propto M^3$, whereas one observes $\eta \propto M^{3.4}$; the theoretical relaxation spectrum is too close to a single exponential. The tube model has also been applied by Doi and Edwards (1986) to rigid rodlike polymers in concentrated solution.

Remarkably, efforts to produce an approximate constitutive equation from this theory, generally produce a variant of the $KBKZ$ model (see for example, Currie 1982, Larson 1988).

[8]Kunihiro Osaki was born in October 1938 in Matsuyama, Japan. His academic career has been associated almost exclusively with Kyoto University, although the two years he spent at the University of Wisconsin, Madison between 1969 and 1971 were influential in moulding his future research interests in polymer rheology. Osaki obtained an MEng at Kyoto University in 1963 and a DEng in Industrial Chemistry in 1968. He became a Professor in the Institute of Chemical Research in 1988. He is currently President of the Japanese Society of Rheology.

[9]Pierre Gilles De Gennes was born in Paris in 1932. Between 1951 and 1955, he was educated at the Ecole Normale Supérieure. He then moved to the University of California at Berkeley as a Postdoctoral Fellow. He returned to France in 1961 to a post at the University of Orsay. He is now attached to the Laboratoire de Physique de la Matiére Condensée, Collège de France. De Gennes is now regarded as one of the leading French scientists of his generation, and he was awarded the Nobel Prize for Physics in 1991. He is a Foreign Member of the Royal Society of London and of the Australian Academy of Science.

In 1987, **Bird** et al published their unique book (*Vol 2 "Dynamics of polymeric liquids; kinetic theory"*), which contains a survey of bead-spring chains, bead-rod and other mechanical models of polymers. A complex phase-space theory is also given; here, not only are the position vectors of the beads considered, but also their velocities. For concentrated solutions and melts, they present an alternative to the Doi-Edwards theory; they say (p 309) "...we do not follow the Doi-Edwards theory, but rather a more systematic kinetic theory approach that is closely patterned after the dilute solution theories". Since the theory uses bead-rod chains, the polymer is not capable of a rubber-like response to rapid transient motions, and the results are somewhat unrealistic; in other respects the theory improves on the Doi-Edwards results (e.g. the steady viscosity function is more acceptable).

The two theories which we have presented in this section show that no complete solution to structure and/or a constitutive equation has yet arisen. New concepts, for example the "double reptation" theory of Graessley (Wasserman and Graessley 1996), continue to appear.

5.6. Suspension Rheology

Suspensions or dispersions of particles in a fluid, are, and always have been, of great importance; blood, ink and paint being some obvious examples. Suspensions of small particles are often termed 'colloids'. Thomas Graham (1805-1869) derived this neo-Greek word from the word for glue. By the 19th century, rubber latex was known; monodisperse lattices became available in the 1950's.

The beginning of our science is the discovery of Brownian Motion; to avoid settling out, the Brownian "forces" must counterbalance gravity; for a particle of radius a, we require $a^4\Delta\rho g/kT < 1$,where $\Delta\rho$ is the density difference between sphere and suspending fluid, and g is gravitational acceleration. Migration of particles is induced by shearing (Leighton and Acrivos 1987), and in many, but not all cases, we can ignore inertia.

The mechanics of suspensions therefore treats identifiable particles moving under the combined effect of Brownian and flow forces. The shapes, stiffness and sizes of the particles, whether or not they are charged, whether or not external electric fields (electrorheology) are present, make the subject an extremely large and growing area of study. For an extensive discussion of the simplest cases, hard spherical particles in solution, Russel[10] et al (1989) may be consulted.

We have already referred to the Einstein (1905) formula (5.1), which applies only to very dilute suspensions ($c \lesssim 0.03$). Interaction between spheres was computed by Batchelor (1977), who found

$$\frac{\eta}{\eta_s} = 1 + 2.5c + 6.2c^2, \tag{5.20}$$

which is applicable for $c < 0.1$. However, the coefficient 6.2 holds only for shearing;

[10] William Bailey Russel was born in Corpus Christi, Texas in 1945. He obtained a BA and MChE at Rice University, before moving to Stanford University, where he obtained a PhD degree in 1973. He joined the Chemical Engineering Department of Princeton University in 1974, where he has helped to create an international centre of excellence in the rheology of colloidal systems. He was elected to the National Academy of Engineering in 1992.

it was found to be 7.6 for elongational flows (Batchelor 1974). Thus, the flow already becomes non-simple (in the sense that shear and elongation are different) at quite low concentrations. Ball and Richmond (1980) used a "self-consistent" field theory to produce the result:

$$\frac{\eta}{\eta_s} = \exp(2.5c) \approx 1 + 2.5c + 3.125c^2, \tag{5.21}$$

which does not reproduce the Batchelor (1974) result at order c^2. Other approaches are by Lundgren (1971), who found the viscosity ratio to be $(1 - 2.5c)^{-1}$ and by Onoda and Liniger (1990), who derived the formula:

$$\frac{\eta}{\eta_s} = (1 - c_m)^{-(\eta)c_m}, \tag{5.22}$$

where c_m is the maximum packing fraction, usually 0.63-0.64. Equation (5.22) is the Krieger[11]-Dougherty (1959) equation, which can be used for various shaped particles. For "hairy" particles, one needs to adjust the radius a suitably (Russel et al 1989).

Experiment shows that for $c \tilde{>} 0.3$, the viscosity becomes sensitive to shear rate. Krieger (1972) has used a group:

$$\tau_r = \sigma a^3/kT, \tag{5.23}$$

to explain this, where τ_r is a reduced shear stress. One eventually finds

$$\frac{\eta - \eta_\infty}{\eta_0 - \eta_\infty} = (1 + \sigma/\tau_c)^{-1}, \tag{5.24}$$

as a description of shear-thinning.

For softer spheres, Mewis[12] et al (1989) obtained

$$\eta_r = \eta_{+\infty} + \frac{\eta_{r0} - \eta_{r\infty}}{1 + (b\tau_r)^m}, \tag{5.25}$$

where τ_r is as above.

Bossis and Brady[13] (1989) developed (two-dimensional) Stokesian dynamics to study suspension mechanics, using computer simulations. Visscher and Heyes (1994) used a 3-D

---

[11] Irvin Mitchell Krieger obtained a BS in Industrial Chemistry at Case Institute of Technology in 1944 and an MS in Physical Chemistry in 1948 at the same Institute. He then worked under the supervision of Peter Debye at Cornell University and obtained a PhD in 1951. Krieger had renewed his association with Case in 1949 and he was to remain there until his retirement in1988. From 1982 to 1988, he was Director of the Centre for Adhesives, Sealants and Coatings. He was awarded the Bingham medal in 1988.

[12] Joannes J F (Jan) Mewis was born in Borgerhout, Belgium in 1938. He studied at the Katholieke Universiteit, Leuven, and obtained a Doctorate in Engineering in 1967. Apart from visiting positions at the University of Delaware and Princeton University, Mewis has spent the whole of his academic career at K. U., Leuven, helping to create a European centre of expertise in colloidal rheology. He was co-Chairman of the 11th International Congress on Rheology, held in Brussels in 1992.

[13] John Francis Brady was born in 1954. He obtained a BS in Chemical Engineering at the University of Pennsylvania in 1975, before moving to Stanford University, where he was awarded an MS in 1977 and a PhD in 1981. He held the post of Assistant Professor at M.I.T. from 1981 to 1985, the year he received the Presidential Young Investigator Award. He then moved to the California Institute of Technology. He is an Associate Editor of the Journal of Fluid Mechanics.

version and were able to show the formulation of strings of particles. Kim and Karrila (1991) used "microhydrodynamics" to study general flows.

Most of the above calculations assume that all particles are of the same size. However, the introduction of a size distribution of particles results in some useful effects. For example, considering a given volume fraction in a monodisperse and a solution of two sizes of particles, it has been found that the monodisperse solution has a higher viscosity than the bidisperse one. At higher concentrations, a more than tenfold viscosity reduction can be achieved (see Barnes 1989).

At high particle concentrations, shear thickening can occur at high shear stresses. Reynolds (1885) discovered the "dilatancy" of sand samples when deformed; by measuring the amount of water between the sand grains, he deduced that deformation increased the volume between the grains; he was led to do the experiment by observing the dry patches under footprints when walking on the beach. Similarly, jamming of the moving particles appears to account for shear thickening; the relevant parameter (Brady and Bossis 1988; Bossis and Brady 1989) is a Péclet number P_e or dimensionless shear $\eta_S\dot{\gamma}a^3/kT$, where η_S is the solvent viscosity, and the particle size is a. Values of P_e of $\sim 10^3$ lead to shear thickening. Although one expects normal stresses in suspensions of spheres, and these are predicted theoretically, measurements are rare.

Thus far, we have discussed spherical particles, but some of the most useful and curious effects occur in suspensions of non-spherical particles, especially long fibres. A more extensive discussion is given in the book by Huilgol and Phan-Thien (1997).

The study by Jeffrey (1922) of the orbits of elongated particles has been fundamental to many later studies, including those involving fibres in solution. An extensive discussion is that of Goldsmith and Mason (1957); see also the review of Frisch and Simha (1956).

Brenner (1974) made an extensive study of rigid spheroids in dilute suspensions. In this non-spherical case, the particle orientation depends on the balance between Brownian diffusion and hydrodynamic forces; again a Péclet number is the governing factor. Hinch[14] and Leal[15] (1975) studied dilute solutions of non-spherical particles and deduced normal stress ratios.

A great deal of interest lies in the use of elongated rod-like fibres of large length/diameter ratio as a reinforcement for plastics. In the forming process, when the polymer matrix is fluid, Folgar and Tucker (1984) produced an approximate constitutive equation for these flows; in such cases, Brownian motion forces are often negligible in comparison with flow forces.

[14]Edward John Hinch was born in Peterborough, England in March 1947. His academic career has been centred on Trinity College Cambridge. In 1973, he obtained a PhD degree there, working under the supervision of Professor G K Batchelor. He became a Fellow of the College in 1971. Currently, he is University Reader in Fluid Mechanics. In 1997, he was elected to a Fellowship of the Royal Society. John Hinch is an Applied Mathematician in the best Cambridge tradition, combining deep physical insight with mathematical skills.

[15]L Gary Leal was born in Bellingham, Washington in March 1943. He received his early education at the University of Washington, before moving to Stanford University, where he obtained an MS in Chemical Engineering in 1968 and a PhD in 1969. After a year spent as an NSF Postdoctoral Fellow at Cambridge University, Leal took up a post at the California Institute of Technology. He moved to the University of California, Santa Barbara, in 1989 as Chairman and Professor of Chemical Engineering. He was elected to the National Academy of Engineering in 1987.

Batchelor (1971) studied the elongational flow of a set of parallel fibres in solution and showed how to compute the stresses arising in such a flow. As the length/diameter ratio of the fibres increases, so the number of particles needed to maintain a dilute solution decreases. A single rod of length L, in a solution of n particles per unit volume, goes from the dilute stage to the semi-dilute stage when $nL^3 \geq 30$. In terms of a fibre diameter d, the volume fraction c becomes $\pi d^2 Ln/4$ and the dilute solution regime ceases when $c \approx 30\pi d^2/4L^2 \approx 24(d/L)^2$. This corresponds to less than 0.25% for an L/d of 100. For higher values of c, a concentrated isotropic regime sets in when $n \approx (dL^2)^{-1}$. Finally, a kind of 'liquid crystalline' state sets in at extreme concentrations. Doi and Edwards (1986) have used a 'cage' model for the semi-dilute regime.

While the above text sketches the cases of interest in mechanical suspension theory in increasing order of difficulty, little account has been given of electrical forces between particles (Russel et al 1989). These introduce another dimension of difficulty to the field. In addition to the ionic forces present in the above studies, there has also been a recent revival of interest in electrorheology, where a strong external electric field is applied to the fluid, thereby inducing "solidification" of the material, enabling large shear stresses to be applied with small or negligible deformation. These materials and effects were first described by Winslow (see, for example Winslow 1947, 1949). In the 1949 paper, he states "A series of experiments begun in 1939 has established that certain kinds of particles suspended in low viscosity oils tend to form an oil-occluded fibrous mass when acted upon by an electric field. The effect has been found intrinsically reversible under action of shear or, in the case of very fluid suspensions, by kinetic agitation alone. The induced shear resistance S, i.e. the increase above that for the de-energized fluid, is found to be nearly a parabolic function of the applied potential V".

In the same paper, Winslow also refers to the analogous *magnetically* induced fibration of ferromagnetic particles in fluid suspension (see, also, Rabinow 1948).

The scale of the ER and MR effects can be quite dramatic, with viscosity increases of several orders of magnitude for electric (and magnetic) fields of modest strength.

Subsequent attempts to develop mechanisms and models for the ER effect are fully discussed in a recent review article by Parthasarathy and Klingenberg (1996), which contains over 150 references. Interestingly, although the obvious applications of ER and MR fluids were openly discussed as early as 1948 (see, for example, Rabinow 1948) and were apparently under consideration at the US National Bureau of Standards before that time, we know of no *successful* commercial exploitation of the effect (cf. Parthasarathy and Klingenberg 1996).

Finally, a most important area of rheology is the study of cracking and mechanical failure; for polymers see the work of T L Smith and others in the book by Ferry (1970); for a general study of the field see the book by Atkins and Mai (1985), where failure of many types of materials is discussed.

It is clear that there is a continued interplay and exchange between continuum rheology and microstructural studies, verging towards polymer and materials science, themselves vast fields of research activity. Regrettably perhaps, we have to limit our History in scope, and we have therefore truncated some aspects of this interface (e.g. crystallization and chemical reactions)) in this chapter. Certain fields, for example dilute solution theory, are in any case making rapid strides by computational studies, and some of these studies are reported in Chapter 8.

5.7. Werner Kuhn

Werner Kuhn, a Swiss physical chemist, was born at Maur, near Zürich, on 6 February 1899, the son of a church pastor. He died at Basel on 27 August 1963.

Kuhn earned his chemical engineering degree at the Eidgenössiche Technische Hochschule (ETH) in Zürich, and then proceeded to a doctorate (1924) on the photochemical decomposition of ammonia at the University of Zürich. He went to work at Neils Bohr's Institute in Copenhagen as a Fellow of the Rockefeller Foundation, and in 1925 he published a soon-to-be famous formula in quantum mechanics: the f-summation theorem, which deals with electric moments of atoms in transition between energy levels. He also met his future wife in Denmark.

In 1927, Kuhn obtained his Habilitation (qualification to lecture) at Zürich and afterwards went to work with K Freudenberg in Heidelberg, where he devised a model of natural optical activity. Three years later, he became Professor Extraordinarius at the Technische Hochschule in Karlsruhe. It was here that he produced many of his early papers on macromolecular behaviour that is of rheological interest, besides continuing his optical work.

Kuhn turned his attention increasingly to macromolecules as we discuss in the text, assuming rod-shaped molecules at first. These early rigid dumbbell models did not agree with the results of Staudinger on the viscosity of solutions, and Kuhn thereupon discovered that the molecules must have the form of a coiled chain. By using this insight and looking at the conversion of chemical to mechanical energy in such systems, he came to some conclusions concerning the operation of muscles.

Kuhn went to Kiel as Professor Ordinarius in 1936 and finally returned to Switzerland as Professor Ordinarius at Basel in 1939. There, he used a countercurrent exchanger to separate heavy water, and the exchanger theory enabled him to understand the mechanism of urine concentration in the kidney and also how high pressure was obtained in the air bladders of fish.

His wide-ranging research earned him honorary doctorates and medals from Kiel, Heidelberg and Bologna. He is described as a wise, calm person in his obituary (Feitknecht 1963).

Kuhn was rector of the University of Basel 1955-6, and was President of IUPAC from 1957-61; his death was unexpected. His call to Basel in 1939 was, for him and his family, a deliverance from political pressure in Germany. He published over 300 papers, and produced, as we have described, some extraordinary new ideas.

His biographer says "Characteristic of all his work is the clear formulation on strong physical foundations of the problems, their exact solution with the necessary mathematical equipment, and the proving of theoretical conclusions through meaningful experiments".

5.8. R Byron Bird

Robert (Bob) Byron Bird was born in Bryan, Texas, on February 5th, 1924. At the time, his father was a Professor of Civil Engineering at Texas A and M University. Shortly after, the family moved back to their roots in Iowa.

Bob Bird's initial undergraduate training in Chemical Engineering began at the University of Maryland (1941-43); it was interrupted by war service and was finally completed in 1947 with a BS from the University of Illinois. A PhD in Chemistry from the University of Wisconsin followed in 1950.

Following a year's postdoctoral work at the University of Amsterdam, Bird returned to the University of Wisconsin, Madison, as Project Associate in Chemistry. He then spent the 1951-52 session as Assistant Professor of Chemistry at Cornell University and the summer of 1952 as Project Chemist at the DuPont Experimental Station. In 1953, he returned to U W Madison, where he has remained ever since, except that is for frequent sabbaticals at Universities in foreign countries, where he has been able to exploit his extraordinary gifts as a linguist. His current position is Vilas Professor Emeritus in the Chemical Engineering Department.

It was largely due to Bird's enthusiasm, expertise and brilliance as a communicator that U W, Madison became an international centre of excellence in rheology. He is one of four Bingham medalists to grace the Rheology Research Centre in post-war years, the others being C F Curtiss, J D Ferry and A S Lodge. Amongst many other honours, Bob Bird was elected a Member of the National Academy of Engineering in 1969 and a Member of the National Academy of Sciences in 1989. He received the National Medal of Science from President Ronald Reagan in 1987.

Bob Bird was introduced to rheology during his summer engagement at DuPont and his subsequent research has covered many strands of the subject. However, he is probably best known for his work on the kinetic theory of polymeric liquids. His books on *"The dynamics of polymeric liquids"* (DPL) have become standard texts, particularly Vol I, *"Fluid Mechanics"* (Bird et al 1987a), and they remain important sources of reference to other workers in the field. These were written in collaboration with R C Armstrong, O Hassager and C F Curtiss. The main motivation for the writing of DPL came from Bob Armstrong and Ole Hassager, who were at the time graduate students at U W Madison. They simply made it known that they would enjoy the task of transforming Bird's excellent lecture notes into a fully-fledged book. On completion of the first draft, it was decided to split DPL into two volumes and Chuck Curtiss was invited to participate in the volume on Kinetic Theory (Bird et al 1987b).

Bob Bird's earlier book on *"Transport phenomena"* has been, if anything, even more influential. This was motivated by his short stay at DuPont and involved collaboration with two Madison colleagues, Ed Lightfoot and Warren Stewart (Bird et al 1960). Incredibly, the book in all its various printings and translations has now sold well in excess of 200,000 copies.

The linguistic abilities of Bird are attested to by his book for English speakers on learning Dutch, and in his two co-authored books on Japanese studies. He is also fluent in Mandarin and, of course, the more common European languages. On another front, his passion for arduous canoe trips in Canada is well-known.

Bob Bird is one of the most feted and well known scientists in the US. He is very hospitable and a charming host.

5.9. Hanswalter Giesekus

Hanswalter Giesekus was born on January 4th, 1922 at Hückeswagen, a small town in the Bergisches Land near Remscheid. His childhood years were overshadowed, first by the world economic crisis, and then by the rise of Naziism. His parents were committed Christians and some persecution resulted, but this was not severe. In due course, Giesekus also embraced the Christian faith.

During his school education, Giesekus had wide interests ranging from literature, history, art and music (he learnt to play the violin fairly well) to mathematics and physics. The prevailing ideological climate strongly suggested a concentration on physics and, after a few months of pre-military service spent partly in France, he went, in the autumn of 1940, to the University of Göttingen. However, after one year's study, he was 'called up' to serve in an air intelligence division, his aptitude for physics being an obvious factor in the decision of the relevant authorities.

From 1943, Giesekus worked in a newly established Institute in Freiburg-im-Breisgau.

During a bombing raid by the British Air Force on the evening of November 24th, 1944, the home where Giesekus was residing was destroyed. He, himself, had a miraculous escape, being on his way to a student bible class at the time!

After the war, in the autumn of 1945, Giesekus' non-involvement in Nazi activities meant that he was immediately able to renew his studies at the University of Göttingen and, notwithstanding the postwar shortages, he obtained his diploma (equivalent to a

'Masters' degree) in the summer of 1948 and then his doctorate in May 1950 in the field of solid-state physics.

Plans to continue research work as a University Assistant Professor did not work out and, in July 1950, Giesekus joined Bayer A.G., based first in an Engineering Department at Dormagen (north of Cologne) before moving to Leverkusen in 1953. At this time, Giesekus became affected by the 'rheology bug' through contact with a colleague, Dr J Pawlowski, one of the main attractions being the possibility of applying methods of modern physics, particularly invariance properties and perturbation theory, to a new class of problems. This was the start of an immensely influential career in rheology and Hanswalter Giesekus soon became one of the leading international experts in the field (see Giesekus 1961a,b,c, 1962, 1963 for examples of his early research in rheology).

Giesekus has been equally at home in micro and macrorheology, in theoretical and experimental rheology, in esoteric and practical pursuits. The breadth and depth of his work in all these aspects has been well summarized in an appreciation by Professor Henning Winter (1989). The accompanying list of references indicates that most of the early papers were published in German and these must have kept German translators busy, as many with an insufficient grasp of that language have realized the value of his scientific works. It is readily admitted by many of his admirers, with an interesting mixture of frustration and admiration, that, after having undertaken a piece of research, they have realized that Giesekus had either published a similar piece of work years before or at least had written something of significant relevance to the work in hand (see, for example, Walters 1989).

Giesekus was the Executive Editor of Rheologica Acta for fourteen years (see Chapter 3) and in that capacity he handled matters with meticulous care; he was suitably firm but always sympathetic in handling papers submitted to the journal and he was forever anxious to avoid injustices arising from any kind of prejudice.

In 1970, Giesekus moved to the newly instituted University of Dortmund as Professor of Fluid Mechanics in the Department of Chemical Engineering. There, he had to build up a research group from virtually nothing. He formally retired in 1987, although he continued lecturing and, of course, researching. In 1990, Professor Giesekus was awarded a Gold Medal by the British Society of Rheology and, in 1994, his long anticipated book *"Phänomenologische rheologie"* was published by Springer.

In 1955, Hanswalter Giesekus married Hanna (Hope) and they have six children, five boys and one girl. Outside his professional work, his Christian faith has consumed much of his time and interests. He is associated with the 'Brethren' movement, which has no formal clergy and the preaching is in the hands of laymen. Giesekus has been very active in this respect and he has also lectured on Christian topics on Transworld Radio. He has taken particular interest in and has written extensively on the life of the French scientist and Christian philosopher, Blaise Pascal (see, for example, Giesekus 1997).

It is difficult to overestimate the impact that Professor Hanswalter Giesekus has had on post war rheology. His own account of his rheological pilgrimage is captured in his Gold Medal address to the British Society of Rheology (Giesekus 1990).

Chapter 6

Rheometry Beyond Viscosity

6.1. Early Measurements of the Normal-Stress Differences in Steady Simple Shear Flow

In this section, we are concerned with the existence and measurement of the two normal stress differences $N_1(\dot{\gamma})$ and $N_2(\dot{\gamma})$. Unlike the viscosity function $\eta(\dot{\gamma})$, interest in N_1 and N_2 is relatively recent and is almost exclusively a post Second World War concern. These so called viscometric functions η, N_1, and N_2 are defined in Appendix 1.

In many ways, early normal stress developments mirrored those associated with the rod-climbing Weissenberg effects (see §7.2) and they essentially grew out of the Second World War experiments on the hydrocarbon gels used as flamethrower fuels. The impressive set of pictures contained in the Weissenberg (1947, 1949) papers already anticipated the way the field was developing (see Figs 6.1 and 6.2). So, if an elastic liquid is contained in the annular gap in a Couette apparatus (say) and either the inner or outer cylinder is rotated, various manifestations of the Weissenberg effect can be observed. The *classic* demonstration of the Weissenberg effect is when the liquid climbs a rotating inner cylinder, the outer cylinder being stationary, but other examples are perhaps more relevant in the present context.

Consider, for example, the related torsional-flow experiment when the liquid is contained between two discs (or between a cone and a plate) (see Fig 6.3 for a schematic diagram of the torsional-flow experiment). The bottom disc is rotated, whilst the top one, containing a number of manometer tubes to monitor pressure, is held fixed. The fluid rises as shown, against the expectation of Newtonian fluid mechanics. Concerning priority of discovery, we may quote Nissan (1994) "I am not sure who first noted the pressure under a stationary disc against a parallel rotating one; most probably the London group[1] by a short time ahead of us. But we were the first to use a 'cone-and-disc' version. Weissenberg acknowledged this - if I remember correctly - in his first letter to Nature by giving thanks to us for pointing it out to him. However, and this is important, it was *his* interpretation which proved to be correct".

The first tangible experimental results on the normal stress differences appear to be those contained in the PhD thesis of R J Russell dated February 1946. Russell was working at Imperial College, London, at the time and he specifically acknowledges Weissenberg's

[1]See §7.2 for a full discussion of the various researchers and research groups who were involved.

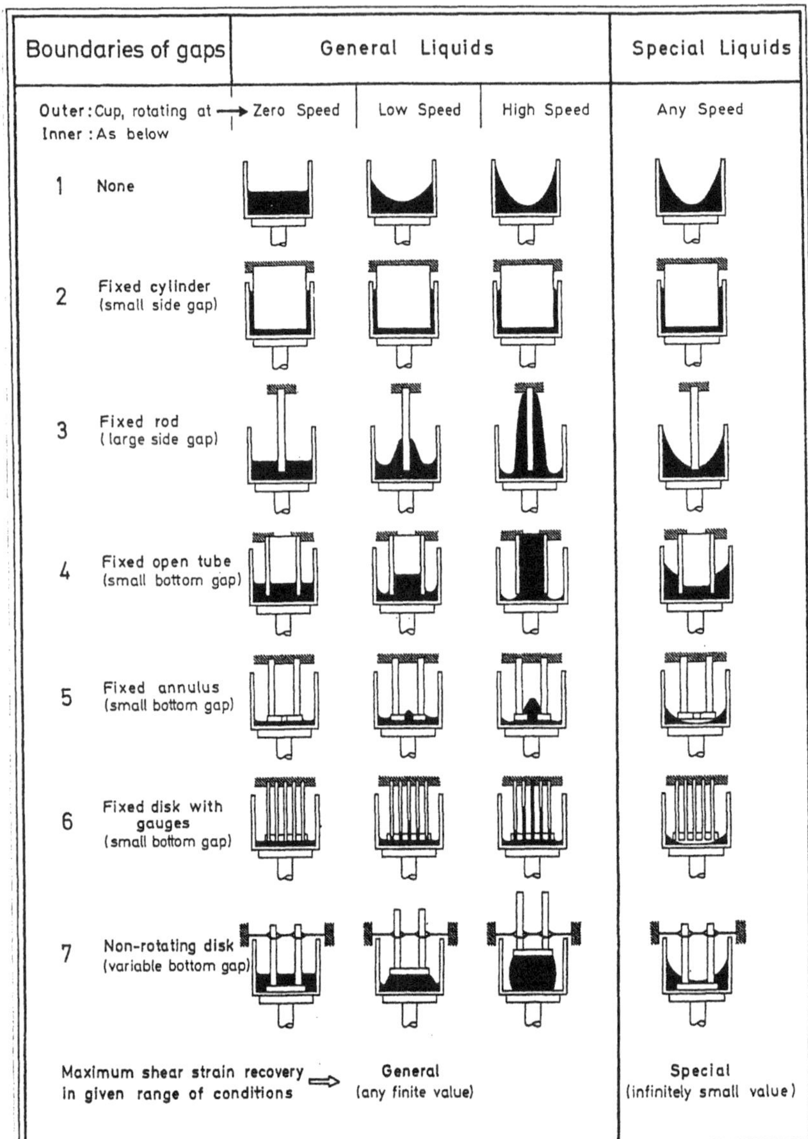

Fig 6.1 The famous drawings entitled "Flow of Liquids under Actions of Stationary Shear in Gaps" taken from the written version of Weissenberg's general lecture at the 1st International Congress on Rheology in Holland (1949). Weissenberg has a convoluted footnote to define 'General Liquids' and 'Special Liquids': "Liquids which are in a certain range 'general' must in the same range be 'non-Newtonian' or 'anomalous', while liquids which are 'special' may be 'Newtonian' or 'normal', but need not be so".

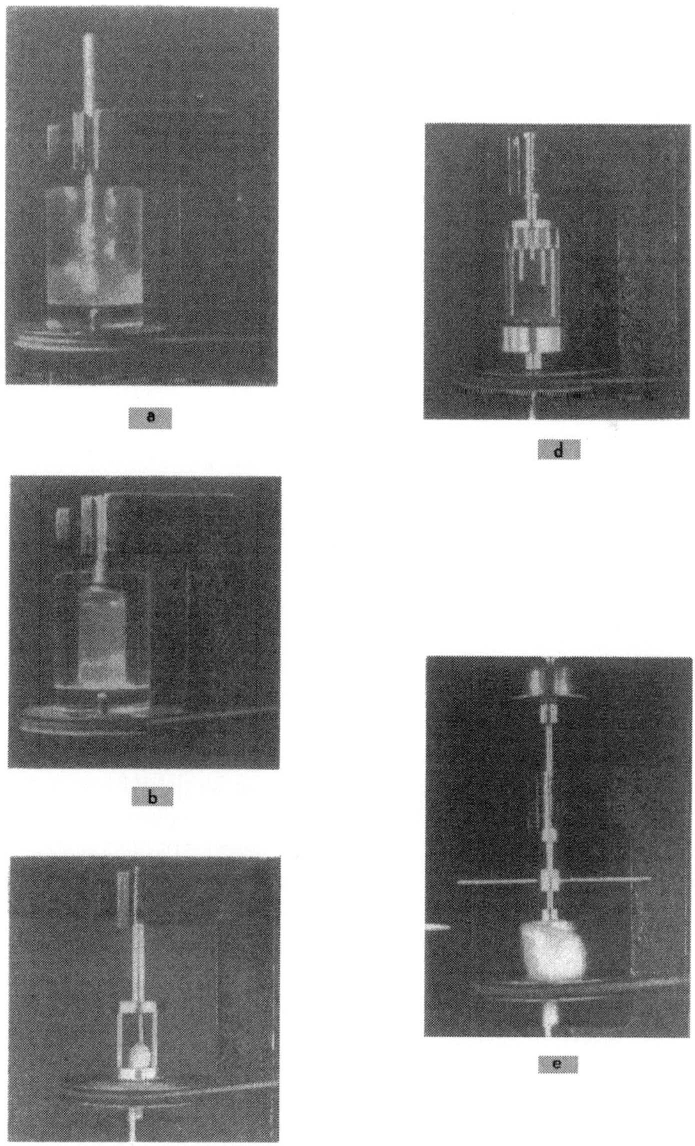

Fig 6.2 Weissenberg's published photographs in the Proceedings of the 1st International Congress on Rheology in Holland (1949). They correspond to some of the schematic drawings in the "High Speed" column of Fig 6.1.

help. The important role of Weissenberg in the overall developments in normal-stress determination is further emphasised by Roberts (1993), who was himself to be a leading player in the developing story. Roberts states "It was not until Karl Weissenberg, who was then Director of Physics Research at the Shirley Institute, Manchester, agreed to act as consultant to the Armament Development Establishment, Langhurst, that the formulation and conduct of experiments was coordinated".

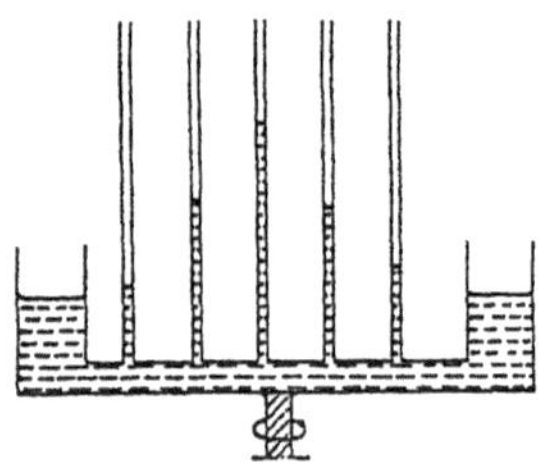

Fig 6.3 Schematic diagram of a torsional-flow apparatus for obtaining normal stress data. (From Coleman et al 1966.)

The cone-and-plate geometry was chosen as the preferred configuration and the resulting instrument, named the Rheogoniometer in 1946, was described by Weissenberg at the 1st International Congress on Rheology in 1948. The use of 'goniometer' was undoubtedly suggested by Weissenberg's earlier work on x-ray diffraction (cf. Chapter 2).

Roberts (1993) states that, later in 1951, it was decided to build a 'quantitative model' to measure the physical properties of gels and to check the validity of the then available theories. Weissenberg agreed to be an advisor and the Model R1 was designed and built at the Armament Development Establishment (ADE), Langhurst. This had a cone-and-plate configuration and the top plate was pierced by a series of manometer tubes. Roberts thus claimed to be the first to make measurements of the distribution of normal pressure and to interpret these in terms of the normal stress differences. As we have already noted, the Birmingham team may have a prior claim to the experimental procedure itself, but Roberts' claim concerning the interpretation of data is probably correct. Roberts found that 'to a first approximation' the results confirmed the Weissenberg hypothesis ($N_2 = 0$), a conclusion that was to be revised or rejected by most later workers in the field with the benefit of improved experimentation (see §6.4).

Modifications and improvements to the Rheogoniometer continued in the following years. Drive through the lower platen and measurement through the upper were replaced by drive and measurement of normal thrust on the lower platen and the measurement of shear stress via the upper. This ultimately led to the first commercial exploitation of the technique. The initial proposal was that Weissenberg would be designated the inventor, Roberts would be designated the designer and Farol Research the maker (cf. Roberts 1993). All in all, nine models (R1-R9) were designed by Roberts at ADE, Langhurst, of which R7 and R8 were the first to be marketed by Farol Research. In 1958, the UK

Ministry of Supply, with changing priorities, ran down its support for rheology research and Roberts effectively disappeared from the scene. Weissenberg remained active however and numerous new developments were to take place in the subsequent years.

It is clear that, in the early commercial developments of the Rheogoniometer, there was a deliberate move from the 'distribution-of-pressure' technique to the simpler integrated normal-force method (cf. Jobling and Roberts 1959). The move was essentially driven by ease-of-experimentation considerations, but little was lost so far as data interpretation was concerned. Specifically, the distribution-of-pressure technique in the cone-and-plate geometry yielded the normal-stress difference combination $N_1 + 2N_2$, through the formula

$$\frac{d\bar{p}}{d(\ell n r)} = -[N_1(\dot{\gamma}) + 2N_2(\dot{\gamma})], \tag{6.1}$$

whereas the total normal force F yielded N_1 directly, through

$$F = \frac{\pi a^2}{2} N_1(\dot{\gamma}). \tag{6.2}$$

In (6.1) and (6.2), $\dot{\gamma}$ is the shear rate, a is the radius of the cone (or plate) and $\bar{p}$ is the pressure on the plate at a radius r.

The corresponding plate-and-plate 'torsional-flow' experiment also played an important role in early normal stress measurements, and interest in this flow can be traced back to Weissenberg's classic set of rod-climbing pictures (see Figs 6.1-6.3). For example, Rivlin (1959), whose careful experimental work with Greensmith (Greensmith and Rivlin 1953) helped to popularize the torsional-flow technique, acknowledges the influence of Weissenberg's 'very beautiful experimental demonstrations'.

This early work of Rivlin initiated a rich vein of rheometrical research in the UK and especially in the US (cf. Rivlin 1971). The British work was located at the Davy-Faraday Laboratory of the Royal Institution and the American work was promoted and funded by the Mellon Institute in Pittsburg, the research programme being overseen by T W DeWitt, assisted by H Markovitz and F J Padden.

Following the publication of the Greensmith and Rivlin 1953 paper, Markovitz constructed a torsional-flow apparatus of essentially the same design. He also adapted it for use in the cone-and-plate mode, and, in the subsequent years, Markovitz and his collaborators made extensive measurements using torsional flow, cone-and-plate flow and also an independent Couette apparatus (see, for example, Markovitz 1957, Markovitz and Williamson 1957). At the time, the Mellon Institute work was influential and it would have remained so, had it not been for the discovery by **Lodge** and his coworkers of the ubiquitous errors involved in using manometers to measure pressure in the case of elastic liquids (see, for example, Broadbent et al 1968). These errors were in part explained by the subsequent theoretical work of Tanner and Pipkin (1969), with the result that all normal-stress measurements using 'holes' to measure pressure had to be viewed as suspect, and the field was left with the mystery of the apparent internal consistency in the work of Markovitz and his collaborators, when hole-pressure errors should have led to large discrepancies (cf. Pritchard 1971). The mystery is still unresolved.

The discovery of the hole-pressure error arose out of the meticulous experiments of Lodge and his collaborators (see, for example, Lodge 1974). Indeed, the Lodge experiments were the epitome of care and precision and they set the standard for all subsequent work in the field. In a comprehensive set of experiments, begun in Manchester, England, and continued at Madison, Wisconsin, following his move to the USA, Lodge and his coworkers (notably A Kaye, J M Broadbent, D G Vale and D M Bancroft) built a number of rheometers with different geometries, including coaxial cylinders, cone-and-plate and parallel plate. Custom-built pressure gauges were employed to measure pressure and a total force facility was also available. Some birefringence work was also undertaken.

In Manchester, all the mechanical instruments were housed in the same room and a comprehensive set of experiments were conducted by Lodge and his coworkers on the same polymer solution. Lodge had an insatiable desire to eradicate, as far as possible, experimental error and extreme care was also taken to control the temperature. The need for vigilance was compounded by the inability to reconcile experimental results from different instruments.

In due course, Lodge postulated that a systematic error associated with the use of holes to measure pressure must be responsible for the anomalies and subsequent research confirmed the hypothesis. Concerned that 'priority of discovery' should be protected, the Manchester team chose the scientific journal 'Nature' as the vehicle for the initial publication (Broadbent et al 1968), the publication time in 1968 being under two months.

Important normal-stress measurements and associated birefringence studies were also performed by Philippoff and his collaborators (see, for example, Philippoff 1957, Brodnyan et al 1957, 1958). However, the associated papers received nothing like the universal acclaim that greeted the corresponding work of Lodge, Markovitz and their coworkers. This lack of enthusiasm may have been due as much to Philippoff's sometimes brusque manner as to any serious deficiencies in the work itself.

For completeness, we briefly draw attention to the following topics of relevance. Capillary and slit methods of normal-stress measurement are discussed in detail in the review article of Boger and Denn (1980); this includes *inter alia* a critique of the exit-pressure work of Han (see also Lodge and de Vargas 1983). A novel instrument, which imposed a total thrust and measured the gap, was discussed by Binding and Walters (1976); it is a normal-stress analogue of the constant-stress rheometers now so often used: Finally, the surprising discovery of *negative* first normal stress differences in liquid crystal systems needs to be noted; see, for example, Kiss and Porter (1978), who seem to have been the first to report the phenomenon.

6.2. Early Theoretical Work on the Normal-Stress Differences

The theoretical work of Reiner (1945b) on inelastic non-Newtonian liquids (which led to the Reiner-Rivlin fluid) certainly indicated the existence of normal stresses in a steady simple shear flow. Reiner at that time was apparently unaware of the classified work on flamethrower gels (see, for example, §7.2) and accordingly an unwarranted constraint on the constitutive functions was imposed by demanding zero normal stresses. Rivlin's (1948a) later work on essentially the same model was more flexible and was viewed by him as being of direct relevance to the growing realization that normal-stress differences were an experimental reality.

The discussion as to whether it is possible to distinguish between inelastic and elastic liquids in a steady simple shear flow has already been discussed in Chapter 4. By the early 50s, it was certainly conceded that the nonlinearities in the constitutive equations for both inelastic and elastic liquids could give rise to normal stress effects (see, for example, Oldroyd 1950, Rivlin 1971), and that the very specific constraint $N_1 = 0$ predicted by the Reiner-Rivlin fluid theory was not appropriate for elastic liquids.

At the same time, Weissenberg (1947), using arguments that many workers in the field found difficult to follow (see, for example, Rivlin 1971) proposed the constraint $N_2 = 0$, which became known as the 'Weissenberg hypothesis'. Much of the early *experimental* work on normal stress measurement seems to have been motivated by and preoccupied with the validity or otherwise of this hypothesis.

From a general continuum-mechanics viewpoint, it became evident that there was no fundamental reason to accept either the Reiner-Rivlin constraint or the Weissenberg hypothesis. Of particular interest is the work of Rivlin(1955, 1956), who noted that in a steady simple shear flow, the Rivlin-Ericksen tensors $\boldsymbol{A}_n$ are all zero for $n \geq 3$. He developed a constitutive equation for such a situation and this involved eight material functions (see §4.4). Thus, for the first time, the possible generality of response in a steady simple shear flow became available.

Coleman and Noll (1959a, 1959b), in an important development and application of simple fluid theory, showed that the response of an elastic liquid in a viscometric flow must be expressible in terms of at most *three* material functions, something that was anticipated by Markovitz and Williamson (1957) and by Criminale et al (1958) and is implied in our notation in Appendix 1.

It has sometimes been argued that the Coleman/Noll analysis is an important improvement on the earlier work of Rivlin (1955, 1956), something that has been vigorously contested by Rivlin himself. So, for example, Truesdell (1965b) in his general lecture at the Fourth International Congress on Rheology states "Here, perhaps, is the place to warn against taking too literally the claim, recently grown common, that the general fluid is "equivalent" to a special Rivlin-Ericksen fluid in viscometric flows. In the narrower sense in which this statement is correct, it tells only half of the true story, namely that the entire flow is governed by the three viscometric functions. To describe viscometric flows, the Rivlin-Ericksen theory with eight material functions is *more* general than necessary".

These sentiments may be contrasted with those of Rivlin (1971): "After Noll's 1958 paper, Coleman and Noll (1959a, 1959b) solved the viscometric flow problem which I had previously solved on the basis of the Rivlin-Ericksen constitutive equation for an incompressible isotropic fluid. The claim was made to greater generality on the ground that they based their calculations on the equation for an incompressible simple fluid with memory. This claim, which was reiterated by Truesdell on a number of occasions, is quite false. Since, for the viscometric flows, all the Rivlin-Ericksen tensors are zero except the first two, the constitutive equation for the simple fluid (our equation 4.8) may for these flows be replaced precisely by equation (17) (our equation 4.5), when full account is taken of material isotropy".

The response of other workers in the field on this issue has varied greatly, depending largely on their discipline and background; it is certainly a fine point. Experimentalists in particular have failed to warm to the nuances involved, preferring to assign them to the domain of the mathematical rheologist. Sufficient to say that in the work of Rivlin (1955, 1956) Markovitz and Williamson (1957), Criminale et al (1958) Coleman and Noll (1959a,

1959b) and Ericksen (1960) there is sufficient generality to underpin all experimental work in the field on isotropic elastic liquids.

The later work of Oldroyd (1965) using his convected-coordinate ideas must be seen as a complementary development, which added little of essential substance to the earlier work of Rivlin and Coleman and Noll.

There is no doubt that, in the early days, theoretical work was motivated by the experimental results that were becoming available. By the early sixties, the tables were turned and the sound theoretical basis underpinning the study of viscometric flows now became the driving force for the growing experimental work.

6.3. Further Commercial Developments on Normal-Stress Measurement

Over the years, the commercial exploitation of the Weissenberg Rheogoniometer has passed through many hands, including, in chronological sequence, Farol Research, Sangamo Controls, Schlumberger, Carrimed and TA. In the early years, the Rheogoniometer was spared any commercial competition. This was to change in the early 70s, with the emergence of "Rheometrics Inc.". The story has its origins at the University of Princeton in the US and involves two young PhD students, Joe Starita and Chris Macosko (see, for example, Macosko and Starita 1971, Starita 1980).

In 1967, Starita completed an MS in Mechanical Engineering working under Bryce Maxwell at Princeton. In the same year, Macosko arrived at Princeton after taking an MSc at Imperial College in the UK. There he was introduced to rheology by Roger Fenner and Gordon Williams; this involved work on a home made cone-and-plate melt rheometer, which used a falling weight to control torque and a mirror to measure rotation.

Both Macosko and Starita were assigned PhD projects which involved the use of the 'Orthogonal Rheometer' that Maxwell had built (see § 6.6). They quickly realised that the existing instrument had serious compliance problems and making this known led to rather heated discussions. However, they decided to redesign the apparatus and persuaded the Department Chairman to purchase a small milling machine, with the argument that, if they failed, it could always be used in the tool shop. They also purchased a wide ranging gear drive and persuaded Triangle Tool Co, where Starita's father worked, to build a stiff X,Y,Z and torque transducer of novel design (US patent 3,693,425).

By early 1969, the new instrument looked promising and Rheometrics was born in early 1970 as a 50/50 partnership between Macosko and Starita. The first instrument was exhibited at the annual US Rheology meeting in Princeton in October 1970. The gears in the original instrument were soon changed to a DC torque motor drive and this provided more flexible speed control.

In 1970, Starita went to work at GE and, six months later, Macosko joined the Faculty of the University of Minnesota. From 1970 to 1972, they both commuted regularly to Triangle Tool Company in New Jersey to supervise the manufacture and testing of the new device.

The so called Mechanical Spectrometer was exhibited at the 1972 International Congress in Lyon, with Imass representing Rheometrics. The response was favourable and everything seemed set for a successful commercial venture. There followed a remarkable series of events.

Starita decided to start another company to make the Mechanical Spectrometer, leaving Macosko with Rheometrics and a substantial debt. Amongst other things, this series of events led Macosko into a deep Christian faith. One of the fruits of this religious conversion was a willingness to turn over all the stock and the patent rights to Starita.

In 1973, Starita left GE to guide Rheometrics into a successful commercial company and hence provide the makers of the Weissenberg Rheogoniometer with their first effective competition. Macosko was hired as a consultant to the firm and still continues in that role. Starita himself held a prominent leadership position in Rheometrics for over two decades, until he sold his interest in the mid nineties.

Although it may not be immediately apparent from the foregoing discussion, the normal stress facility on what became known as the Rheometrics 'Mechanical Spectrometer' essentially grew out of a rheometer (the Orthogonal Rheometer) which was originally constructed to investigate the linear time dependent behaviour of polymeric systems, through such functions as the storage and loss moduli. In the same way, the third commercial player in the area of all-purpose rheometers, namely Bohlin, entered the field through a restricted rheometrical background. Leif Bohlin had worked in the Department of Food Technology at the Chemical Center, University of Lund, Sweden. There, he had developed a stress-relaxation instrument to investigate the long-time stress-relaxation of wheat flour dough. Later, a controlled-strain oscillation instrument was also developed to study coagulation and gelation of various food-related substances. These developments ultimately led to the construction of the Bohlin VOR rheometer during late 1982 and early 1983, the first commercial instrument being delivered in May 1983. Bohlin resigned his University post in 1984 to devote himself full-time to the fledgling company, Bohlin Rheologi AB.

One stunted development in rheometer development concerns the instrument firm Instron. In 1970, R I Tanner became a rheological consultant to the firm with a view to giving advice on a possible rotary rheometer. Development and design progressed favourably, but there was a hiatus of two years following the launch of the Rheometrics' instrument in 1970; one of the main reasons being the possibility of an Instron buy-out of the fledgling Rheometrics Company. However, no agreement was reached and, after two years, Instron resumed development on the rheometer. By the time the 3250 instrument was ready in 1975, Rheometrics had a five years start. Furthermore, the Instron development was moved to the UK and it is hardly surprising that the instrument failed to make inroads into the expanding US rheometer market - a clear example of the business maxim that 'hesitation is deadly'.

6.4. The Second Normal-Stress Difference

The measurement of the *first* normal stress difference N_1 has never been particularly controversial and all reliable experimental data for isotropic elastic liquids show a positive N_1 for all shear rates, with N_1 increasing monotonically with shear rate.

The story concerning the *second* normal stress difference N_2 is far less straightforward and the relevant history is more interesting. This is due to a number of factors. These include the comparative difficulty of making accurate measurements of N_2 and also the crucial importance of the hole-pressure error in many of the early experiments.

The interesting and illuminating part of the story relates to the growing number of papers on the subject in the 60s and early 70s, this being the halcyon period of normal stress measurement. During this period, numerous authors provided compilations of the various contributions to the normal stress literature (see, for example, Ginn and Metzner 1969, Olabisi and Williams 1972, Miller and Christiansen 1972, Pipkin and Tanner 1972) and, in Table 6.1, we include the compilation provided by Walters (1975) for rotational rheometers. There was also the attempt to measure N_2 directly using an annular flow (Hayes and Tanner 1965), which was plagued by hole-pressure errors and gave $N_2 > 0$, and the successful measurements using an open tilted trough (Kuo and Tanner 1974, Keentok et al 1980), which did give $N_2 < 0$ (see Fig 6.4).

The confusion over the sign of N_2 is apparent from Table 6.1, but the chronology is clearly important. The pre-1962 work was not sensitive enough to detect departures from the Weissenberg hypothesis ($N_2 = 0$) and much of the research between 1962 and 1967 was clouded by the popular usage of holes to measure pressure. Generally speaking, methods using holes indicated a positive N_2, whereas those relying on total-force measurements or *flush-mounted* pressure transducers pointed in the direction of a negative N_2. The confusion was resolved once the hole-pressure error was admitted into the analyses and it is now generally agreed that available experimental evidence is consistent with a negative N_2, which is usually much smaller in magnitude than N_1. A word of caution is however in order at this point. Most of the reliable data have been obtained on a limited range of well characterized highly-elastic polymer solutions and it is important to remember that in general there appear to be no fundamental reasons to restrict N_2 to be small and negative. Indeed, recent data, obtained using modern experimental techniques (see, for example, Magda et al 1991, C-S. Lee et al 1992a, Magda and Baek 1994), indicate absolute values of N_2 which range from the very small to over 25% of N_1, depending on the particular polymeric liquid investigated.

6.5. Rheo–Optical Techniques in Normal-Stress Measurement

As an alternative to mechanical methods of determining the normal stress differences, attempts have been made to exploit the rheo-optical properties of some non-Newtonian systems. These are not always appropriate for a variety of reasons (see for example, Janeschitz-Kriegl 1983), but, where they are, birefringence studies can be important in providing normal stress data.

White (1990) has traced the history of rheo-optics, starting from the work of Huygens (1690) in the 17th century, through the studies of J C Maxwell (1853, 1873) in the 19th century, to the expanding interest of the present century.

It appears that the significant modern developments date from the 40s and 50s and the names of Kuhn and Grün (1943) and Stein and Tobolsky (1948) figure in the developing story. In 1956, Lodge showed that the Kuhn and Grün arguments could be extended to a network of entangled chains, and hence were appropriate for polymer melts and solutions. Philippoff and co-workers (see, for example, Brodnyan et al 1957, 1958, Philippoff 1956, 1960a,b, 1961) made important experimental contributions to confirm the growing expectations, setting the stage for the definitive works of Wales (1976) and **Janeschitz-Kriegl** (1983) and the more recent studies of Fuller (see, for example, Fuller 1995).

Table 6.1. Second-normal stress measurements from rotational instruments.

Investigators	*Year*	*Apparatus and method*	*Hole pressure error accommodated*	*Results*
Roberts	1952	Total force and pressure distribution in cone-and-plate flow.	No	$N_2 = 0$
Kotaka, Kurata and Tamura	1959	Pressure distribution and total force in parallel-plate flow.	No	$N_2 = 0$
Markovitz and Brown	1963	Pressure distribution in cone-and-plate and parallel-plate flows.	No	$N_2 > 0$
Adams and Lodge	1964	Pressure distribution in cone-and-plate and parallel-plate flows.	No	$N_2 > 0$
Jackson and Kaye	1966	Total force in extended cone-and-plate flow.	Not applicable	$N_2 < 0$
Kaye, Lodge and Vale	1968	Pressure distribution in Couette, cone-and-plate and parallel-plate flows.	Yes	$N_2 < 0$
Marsh and Pearson	1968	Total force in extended cone-and-plate flow.	Not applicable	$N_2 < 0$
Ginn and Metzner	1969	Total force in cone-and-plate and parallel-plate flows.	Not applicable	$N_2 < 0$
Berry and Batchelor	1968	Total force in cone-and-plate and parallel-plate flows.	Not applicable	$N_2 < 0$
Cowsley	1970	Total force in extended cone-and-plate flow.	Not applicable	$N_2 < 0$
Pritchard	1971	Total force and/or pressure distribution in Couette, cone-and-plate or parallel-plate flows.	Yes	$N_2 < 0$
Miller and Christiansen	1972	Total force and pressure distribution in cone-and-plate flow.	Yes	$N_2 < 0$
Olabisi and Williams	1972	Total force and pressure distribution in cone-and-plate flow.	Yes	$N_2 < 0$
Brindley and Broadbent	1973	Total force and pressure distribution in cone-and-plate flow.	Yes	$N_2 < 0$

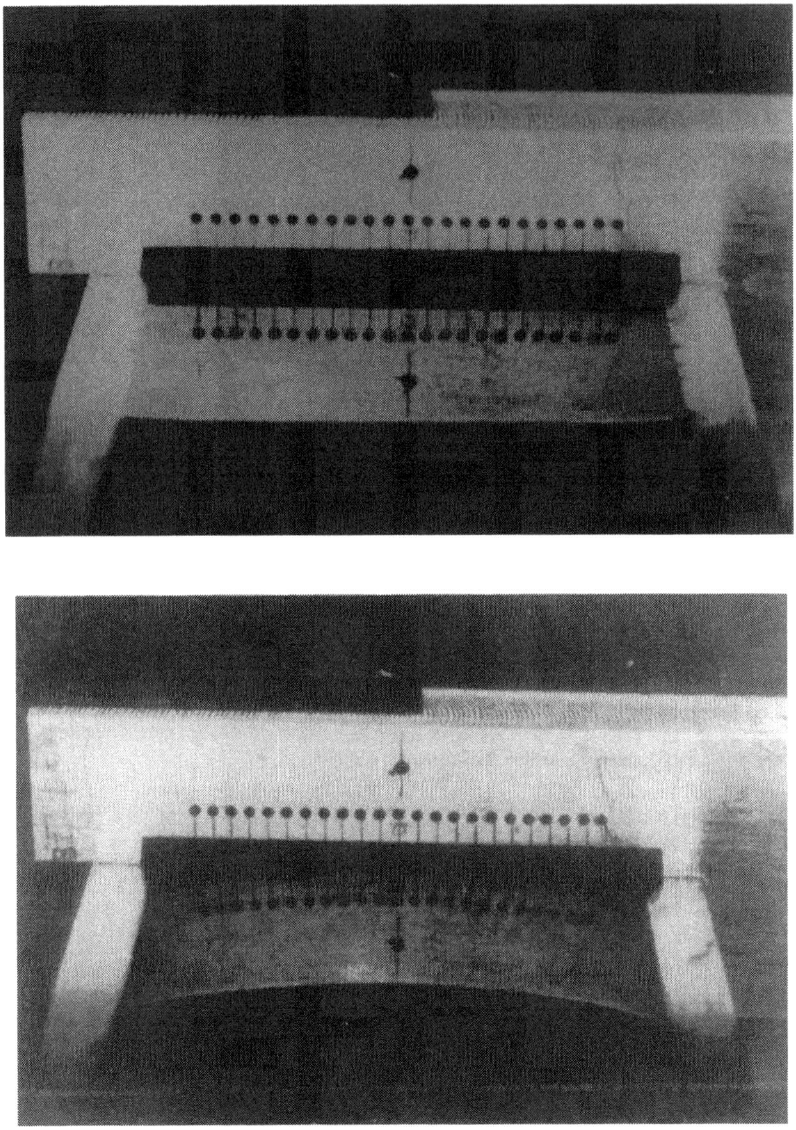

Fig 6.4. The photographs show the reflection of a straight edge in the surface of an open inclined trough of semicircular cross-section, down which fluid is flowing under gravity.The top photograph shows that the Newtonian-fluid surface is flat, filling exactly the semicircular channel. Below, a non-Newtonian fluid bulges up, a consequence of N_2 being negative in this polyethylene oxide solution. (From Keentok et al 1980.)

The important point at issue is the relationship between birefringence and the state of stress, expressed via the so called stress-optical law. For a wide range of conditions, it became apparent that a *linear* stress-optical law holds; i.e. the deviatoric components of the refractive index tensor and the stress tensor are proportional to each other. Furthermore, there is usually no observable time lag between stress and birefringence, although the evidence here is certainly not unequivocal.

Setting the limits on the linear law is clearly an important matter and this is considered in detail in the text of Janeschitz-Kriegl (1983).

Optical rheometry must be viewed as an important and complementary addition to mechanical rheometry; it has enriched the field in those areas where the appropriate techniques have been shown to have utility.

6.6. Linear Viscoelasticity

In Chapters 1 and 2, we introduced the concepts of linear viscoelasticity and emphasised their importance. The discussion in those chapters was essentially theoretical, and we now need to address corresponding *experimental* developments. We do so by first noting that, since the Second World War, experimentalists have introduced alternative functions to interpret viscoelastic behaviour, most notably the storage modulus G' and loss modulus G'' or equivalent functions involving 'viscosity' rather than 'rigidity'. The moduli arise quite naturally in the interpretation of experiments involving simple harmonic shearing at small amplitudes. These new functions, and indeed all the various functions used to interpret linear viscoelastic behaviour, are related by simple formulae which often involve integral transforms (see, for examples, Barnes et al 1989).

White (1990, p126), in tracing the history of experiments to determine the functions of linear viscoelasticity, points to the 1930s as the time when serious small strain studies began to emerge. He refers specifically to the work of Schofield and Scott Blair (1932) on flour doughs and Eisenschitz and Philippoff (1933) and Philippoff (1934a,b) on polymer solutions and polymer melts. In the 1940s, further influential contributions came from Leaderman (1943) and Tobolsky and co-workers (see, for example, Andrews et al 1948). Leaderman, Tobolsky and Philippoff were prominent rheologists in a time of rapid expansion; each of them was awarded the Bingham medal, in 1955, 1956 and 1962, respectively (see Appendix 2).

By the early 50s, attention was focusing on small amplitude oscillatory shear measurements and coaxial-cylinder instruments were built by the UK Courtauld's group at Maidenhead (Oldroyd et al 1951), by the Mellon Institute group in Pittsburg (Markovitz et al 1952) and by Ferry and his collaborators (Smith et al 1949). The first two instruments utilized *angular* oscillations about the axis of the cylinder, while the latter involved a rod oscillating *axially* in a closed tube.

Later, commercial instruments began to appear and dynamic measurements soon became commonplace. These tended to concentrate on the cone-and-plate geometry, the choice being largely governed by the popularity of that geometry in the corresponding steady shear experiments (see §6.1).

An interesting interlude in linear-viscoelasticity testing occurred in the 60s with the advent of a group of rheometers which utilized a novel method to generate an oscillatory flow. In all these instruments, the flow was steady in an Eulerian sense, although the fluid sample was exposed to an unsteady deformation. Typical of these devices were the Orthogonal Rheometer and the Balance Rheometer (see, for example, Walters 1975).

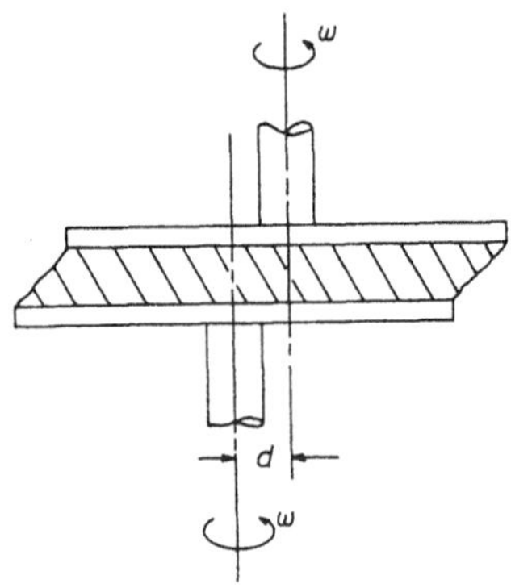

Fig 6.5. Schematic diagram of the Orthogonal Rheometer.

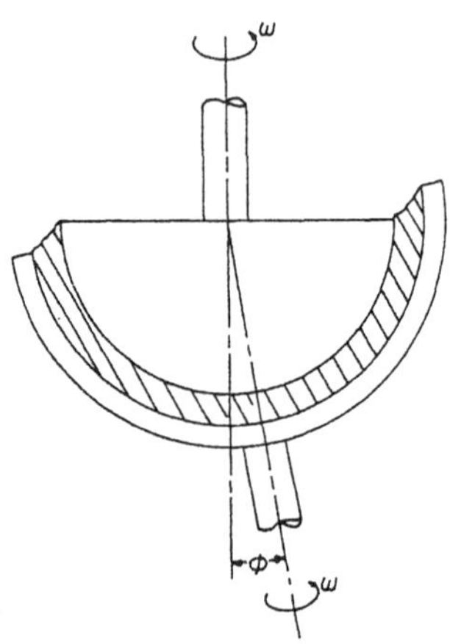

Fig 6.6. Schematic diagram of the Balance Rheometer.

The Orthogonal Rheometer shown schematically in Fig 6.5 was initially introduced by Gent (1960) in a study of crosslinked rubbers. It was later utilized by Mooney (1964) and then popularized by Bryce Maxwell of Princeton (see, for example, Maxwell and Chartoff 1965). This instrument essentially consisted of two flat parallel plates, which rotated with the *same* angular velocity about two axes, which were normal to the plates but not coincident. The test fluid was contained between the plates and the two components of the tangential force on one of the plates were measured, which yielded the dynamic data directly. One of the problems with the Orthogonal Rheometer was that the forces due to G' and G'' tended to deflect the plates, and fairly elaborate computation was needed to ensure accurate results, especially in the case of stiff materials.

In the Kepes Balance Rheometer shown schematically in Fig 6.6 (see, for example, Kepes 1968, Kaeble 1969, Jones and Walters 1969, Marin 1988), the test fluid was contained between two concentric hemispheres and the components of the torque on the inner hemisphere were measured, but the fundamental idea was similar to that of the Orthogonal Rheometer.

Attempts were made to market commercially both the Orthogonal Rheometer and the Balance Rheometer, but, after a brief period of prominence, they lost their appeal, and linear viscoelasticity testing reverted to the more conventional small-amplitude oscillatory-shear techniques, with creep testing favoured for some applications.

One development concerning the *interpretation* of experimental data on polymeric liquids is worthy of comment. This concerns Time-Temperature superposition and the associated 'Method of Reduced Variables'. The development is very much associated with the name of **J D Ferry** (see, for example, Ferry 1970). In order to study the full range of rheological properties for varying temperature (and sometimes pressure and polymer concentration as well), normalized variables are used and, by so doing, a unique curve is obtained whatever the temperature (or pressure or concentration). The method is basically empirical, but is guided by reference to molecular theory. The so-called WLF equation is often involved in applications of this procedure, the initials referring to M L Williams, R F Landel and J D Ferry (see, for example, Williams et al 1955 and Bird et al 1987a).

6.7. Extensional Viscosity

Extensional flow and the associated extensional viscosity η_E defined in Appendix 1 play an increasingly important role in current rheological research, especially as that relates to polymeric fluids. In many ways, the subject may be viewed as peculiarly *rheological* in its appeal, since Newtonian liquids do not provide any significant challenges in the area. In particular, it is well known that, for Newtonian liquids, the extensional viscosity is simply three times the usual shear viscosity, a result commonly attributed to Trouton (1906).

Recent interest in extensional viscosity arises in part from the realization that, for polymeric liquids, the ratio of the extensional viscosity to the shear viscosity (now commonly referred to as the Trouton ratio) can be significantly higher than 3, implying that in relative terms these liquids, particularly *dilute* polymer solutions, often show much more resistance to an extensional flow than would be anticipated from their shear viscosity response. A dramatic demonstration which can be related to this exceptional resistance to an extensional deformation is given in Fig 6.7 (see also §7.4).

Early interest in extensional flow can be traced to the classic paper of Trouton (1906), who, as part of a general study of the viscosity of pitch and similar very viscous materials, investigated a number of methods to determine η_E, (which he called 'the coefficient of viscous traction'). Trouton also studied the subject theoretically and correctly deduced the factor 3 between η_E and η. For some unaccountable reason, Trouton's theoretical work was not generally appreciated or acknowledged, and in the decades that followed subsequent papers contained theoretical errors that could and should have been avoided in view of Trouton's analysis (see, for example, the discussions in Petrie 1979 and Petrie and Dealy 1980). A further illustration of this oversight is provided by the fact that, as late as 1941, Wiley attributed the factor 3 to Burgers (1935) rather than Trouton.

The sixty years or so that followed the publication of Trouton's (1906) paper were not noted for seminal research on extensional flow and few publications are worthy of comment. In 1943, Scott Blair did highlight the experimental techniques that were then currently available, and Andrade and Chalmers (1932) and Nitschmann and Schrade (1948) introduced techniques which were the precursors to a number of later experimental developments[2]. But these were isolated publications in a barren period. It was not until the middle to late 60s that extensional flow suddenly blossomed as a subject of significant rheological interest (see, for example, Walters and Barnes 1980). The impetus came largely from the realization that the Trouton ratio for polymeric liquids could be very high indeed and certainly significantly greater than the 3 for Newtonian liquids.

The basic technique for extensional viscosity measurement is the homogeneous stretching method, which is shown schematically in Fig 6.8. In traditional tensile testing machines, the movable block has a constant velocity. This does not generate a constant extensional strain rate. To generate a *constant* rate of strain, it is necessary for the velocity of the movable block to vary exponentially with time. An alternative procedure is to provide a constant *stress*, which is brought about by applying a force to the movable block, which decreases in proportion to the cross section of the extending specimen.

The early papers in the period of growing research activity were written by industrialists. The paper of Ballman (1965) of Monsanto in the USA, using a constant strain-rate device, is seen as heralding the renaissance (or more correctly "the birth") of interest in extensional flow and this was followed by papers on a constant *stress* device by Cogswell (1968, 1969) of ICI in the UK. At the same time, BASF in Germany provided substantial support for research in the subject and the names of J Meissner, H M Laun[3] and H Münstedt[4] figure prominently in the expanding literature of the period.[5]

---

[2]In the case of Nitschmann and Schrade's 'spinning' work, it is probably more accurate to say "*re*introduce", since Trouton (1906) had used the spinning technique in some of his experiments.

[3]Hans Martin Laun was born in 1944 in Heilbronn, North Germany. He began his university education in physics at the University of Stuttgart, before moving to the University of Ulm, where he obtained a doctorate in 1974. He then joined BASF, Ludwigshafen, to commence a distinguished career as an industrial polymer rheologist. Since 1995, he has been group leader in the "Rheology, Materials Property" department. Laun has the distinction of serving on the Editorial Boards of the three premier journals in the field.

[4]Helmut Münstedt was born near Hannover, Germany in 1941. His university education in Physics began at the Technical University, Braunschweig, and continued at the Technical University, Stuttgart, where he obtained a doctorate in 1971. After a year spent as a Post-doctoral Fellow at the City University, London, Münstedt returned to Germany to work at the BASF Polymer Research Laboratory in Ludwigshafen. 1989 saw his appointment to a professorship at the Technical University of Berlin and 1991 his appointment as Director of the "Technical Service of Low Density Polyethylenes" department at BASF. Since 1993, he has been Head of the Institute of Polymer Materials within the Department of Materials Science at the University of Erlangen-Nürnberg.

[5]In 1981, Meissner, Laun and Münstedt, in association with M H Wagner, were jointly awarded the Annual Award of the British Society of Rheology for their work on extensional flow.

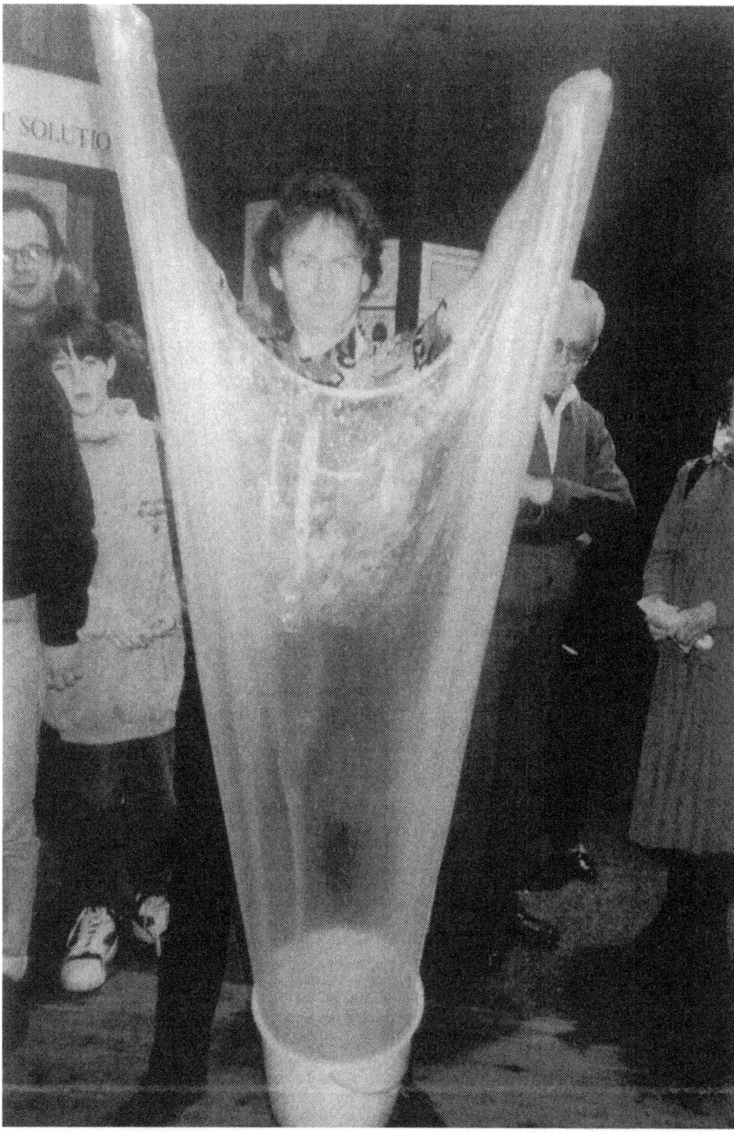

Fig 6.7. A highly elastic liquid is contained in a bucket. Portions of the liquid are clasped in both hands and the hands then raised abruptly. This results in virtually the whole of the contents being extended in the form of a sheet. The exhibitor is Mr Rod Binnington; the occasion is "the Great Australian Science Show" held at the Royal Exchange building, Melbourne, in 1993; the liquid is a 2.2% solution of polyacrylamide in water (17.8%) and glycerine (80%), sometimes known as 'super goop'.

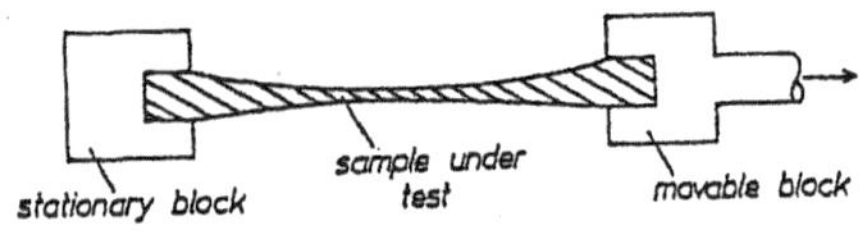

Fig 6.8. The homogeneous stretching method.

The 'rotational clamps' technique of **Meissner** (1969, 1972) was particularly novel and influential. In this, two sets of wheels rotate with constant angular velocity, thus ensuring a constant rate of strain. This period of expanding research interest also saw important contributions from Russian scientists, particularly **Vinogradov** and co-workers (see, for example, Vinogradov and Malkin 1980) and also Münstedt (1975), who further perfected the constant-stress technique.

As the field developed and expanded, it was realized (see, for example, Acierno et al 1974) that polymer melts and *concentrated* polymer solutions had a similar qualitative response in an extensional flow which, although typified by Trouton ratios in excess of 3, was not as excessive as that observed for *dilute* solutions of flexible polymers, where very high Trouton ratios of 1000 or more were not uncommon. This fact helped to motivate the search for techniques to determine η_E for *mobile* polymer solutions. These have included open-syphon flow, adapting the original Fano idea (see §7.4); the Spin Line Rheometer (cf Hudson and Ferguson 1976), based on the basic fibre-spinning technique; flow through a contraction, with Binding (1988) adapting and extending the ideas of Cogswell (1972a, 1972b); stagnation-flow devices (see, for example, Fuller et al 1987) and other even more hybrid experimental methods.

International Workshops in the mid eighties highlighted the difficulties of obtaining meaningful extensional viscosity data for dilute polymer solutions and matters came to a head when a number of laboratories used *different* techniques to provide η_E data for the *same* polymer solutions. The first such exercise, for a constant-viscosity highly-elastic (Boger) fluid provided by T Sridhar of Monash University, Australia, and designated M1, highlighted dramatically the scale of the problem (see Journal of non-Newtonian Fluid Mechanics Vol 35, 2-3, 1990 and cf Fig 6.9). The various contributions to the round-robin exercise were not subject to excessive experimental error and the discrepancies shown in the figure are simply a reflection of the difficulty of generating a true extensional deformation for a sufficiently long time to permit the determination of meaningful η_E data. So the chaos shown in Fig 6.9 is now understood and is assigned to the imperfect kinematics in the various techniques. Attempts are being made to overcome these very severe difficulties, using, in the main, sophisticated adaptations of the spinning technique.

The collaborative ventures in the study of extensional viscosity for mobile polymer solutions matched an earlier enterprise for polymer *melts*, resulting from a series of IUPAC Working Parties on three similar LDPE samples (see for example, Meissner 1975).

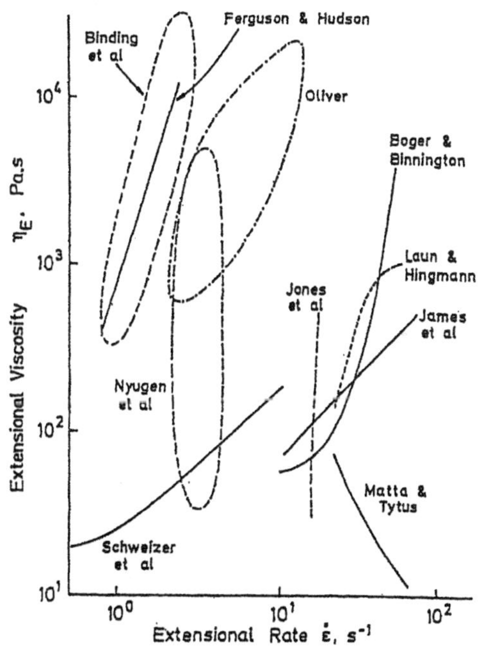

Fig 6.9. Extensional viscosity measurements on the M1 fluid. The lines and envelopes represent data by the various M1 investigators, as published in the special M1 issue of the Journal of non-Newtonian Fluid Mechanics. (From James and Walters 1993.)

On the theoretical side, two influential publications, which are often quoted, are those of Batchelor (1970, 1971), who studied the extensional viscosity characteristics of suspensions of slender particles. A less well known publication of Roscoe (1965) is also worthy of comment. He obtained a theoretical formula for a simple fluid, which may be written in the form:

$$\left.\frac{d\eta_E}{d\dot{\epsilon}}\right|_{\dot{\epsilon}\to 0} = \tfrac{3}{2}\left.\left[\frac{N_1 + 2N_2}{\dot{\gamma}^2}\right]\right|_{\dot{\gamma}\to 0}. \tag{6.3}$$

This relates the extensional viscosity at very small strain rates to the normal stress differences in a shear flow at very low shear rates.

6.8. H Janeschitz-Kriegl

Herman Janeschitz-Kriegl was born in Graz, Austria on October 22nd 1924. The period between February 1943 and the end of the Second World War in April 1945 was taken up with military service. Following the war, Janeschitz-Kriegl studied physical chemistry and theoretical physics and obtained a doctorate in the summer of 1951. He worked for a local pharmaceutical firm in Graz before moving to the Central Laboratory TNO, Delft, in the Netherlands in 1952. In his own words "After a youth, which was very harsh, with finally an immigration in order to obtain a promising job, our generation took advantage of the rapid development gaining impetus in the mid fifties. So, Holland certainly became our second and beloved home country, where our children were born, where we acquired our friends for life". A J Staverman and F Schwarzl were distinguished colleagues and friends during this period.

Janeschitz-Kriegl stayed at TNO until 1968, when he made the short move to take up the post of Professor of Chemistry and Physics of Macromolecular Compounds at Delft University of Technology. In 1978, he returned to Austria to take up a position as Professor of Physical Chemistry at Johannes Kepler University in Linz. In his own words "Our return to Austria was quite unforeseen. It suddenly became discussable, when the Austrian government started a move to bring Austrian scientists home. For me, the greatest advantage was that my prospect of continuing scientific work at advanced age improved tremendously, because of a very advantageous contract. So, my years in Austria became the most productive years of my life. Another advantage is the fact that one can live here more close to nature than in the overpopulated Netherlands. On the other hand, the touch of internationality, so characteristic of the Netherlands, cannot be felt so clearly in Austria". Janeschitz-Kriegl formally retired in September 1995, but he now continues researching as a Professor Emeritus.

Janeschitz-Kriegl is best known for his work on flow birefringence, but there have been other discoveries during his long career, including the invention in 1963 of a probe for temperature measurements in flowing polymer melts and the introduction of a slit rheometer for polymer melts in the same year.

In 1953, Herman Janeschitz-Kriegl married Gertrude Gretitsch; they have four children. The family contains a number of very talented musicians; they sometimes play together as a string quartet. Indeed, chamber music remains Hermann Janeschitz-Kriegl's abiding passion, after his science and his family.

6.9. John D Ferry

John Douglass Ferry was born on May 4th 1912 in Dawson, Yukon Territory, Canada, where his father was a mining engineer. He obtained his Bachelors degree from Stanford University in 1932 and his PhD from the same University in 1935. His doctoral work involved a protracted stay at the University of London, England.

Ferry moved to Harvard University in 1936 and he spent nine years there in various capacities. In 1946, he was appointed Assistant Professor in the Chemistry Department of the University of Wisconsin at Madison. His association with UW, Madison was to be a long and highly successful one. He very quickly moved through the ranks and he had been appointed to a full Professorship by 1947. Later, between 1959 and 1967, Ferry became Chairman of the Chemistry Department and, from 1973 until his official retirement in 1982, he held the position of Farrington Daniels Research Professor.

During an illustrious academic career, Ferry received numerous honours. For example, in 1953 he was awarded the Bingham medal by the Society of Rheology; he was Honorary President of the International Congress on Rheology in Kyoto, Japan, in 1958; he was President of the American Society of Rheology between 1961 and 1963 and, from 1961 to 1963, he also served as Chairman of the International Committee on Rheology.

Madison became an international centre of excellence in rheology and, during his tenure there, Ferry was to supervise over 50 research students, two of whom (T L Smith and D J Plazek) were later to receive the Bingham medal of the S.O.R. Another research student, the late R S Marvin, was to play a prominent administrative role as Secretary of the International Committee on Rheology.

Ferry's classic book *"Viscoelastic properties of polymers"* was first published in 1961. It soon became a standard reference in the area of 'linear viscoelasticity' and a third edition was published in 1980. In addition, over 300 research publications flowed from his pen.

At the height of his powers, John Ferry was a highly successful and influential rheologist. However, he remained unassuming and charming - a true gentleman.

6.10. Joachim Meissner

Joachim Meissner was born in Sehma/Annaberg, Saxony, Germany, in 1929. He obtained his elementary and high school education at Bitterfeld and then, between 1947 and 1949, he worked as an apprentice electrician. His University education stretched from 1949 to 1958 and this was centred at the University of Stuttgart, where he majored in Physics. At Stuttgart, he obtained a Masters degree and then a PhD in metal physics.

In 1958, Meissner moved to BASF (Badische Anilin - and Soda - Fabrik), Ludwigshafen am Rhein, the first two years being spent in the Plastics Engineering Applications Department. He then moved to the Polymer Physics Research Laboratory and, in 1970, he became leader of the polymer melt rheology group. The BASF years were very important for Meissner. They coincided with the surge of international interest in extensional flow and he was to become a major player in the field. His association during those years with H M Laun, H Münstedt and M H Wagner was particularly noteworthy and influential.

In 1974, Meissner moved to the Eidgenössische Technische Hochschule (ETH), Zürich, Switzerland as Professor of Polymer Physics. There, he continued his innovative work on experimental methods to determine the extensional viscosity of polymer melts. The relevant instruments were novel and sophisticated and the Swiss environment and technical backup provided an excellent setting for a number of influential experimental developments. During those years, Meissner was very much sought after as a plenary lecturer at scientific gatherings and his lucid presentations were invariably illustrated by means of films of the highest quality.

Meissner officially retired from his ETH polymer physics post in 1996, but he is now actively involved in a new ETH venture called the Swiss Rheocenter, working closely with his successor, Professor H C Öttinger, and Professor E Windhab.

Joachim Meissner has a unique and sometimes abrupt lecturing style and he often fends off hostile questions with an abrupt manner, which can seem curt and dismissive. But this would be an unfortunate and mistaken impression and a characteristic glint of the eye and amusing quip can be quite disarming. He has many friends in the rheological community and he remains the leading international expert in polymer melt extensional rheology.

6.11. G V Vinogradov

Georgii Vladimirovitch Vinogradov was born into a distinguished family in Moscow on January 24th, 1910. His father, Professor V V Vinogradov, was a famous Soviet scientist and a member of the Russian Academy. He became the personal physician to Joseph Stalin, but in 1952 he was falsely accused of conspiring to murder Soviet leaders, according to Stalin's daughter, Svetlana Alliluyeva (1968). It was only the death of Stalin that saved V V Vinogradov from execution and disgrace, and the family from being sent into exile.

Georgii Vinogradov graduated in chemistry from the Moscow Technological University in 1931, and in 1932 he joined the Red Army, commencing his service in Military Academies. At that time, such a career provided the best means of working as a scientist in relatively good conditions.

The first research papers from the pen of Vinogradov appeared before the outbreak of the Second World War; they were in the area of physical chemistry. During the war, he served in the Soviet Tank Corps, and he was introduced to rheology when he became responsible for the greases used in tank bearings. So began a fruitful period of research into the rheology of greases and friction and wear. This lasted through the 40s and 50s and a highly qualified group of experts was assembled. Like the corresponding problems associated with the normal-stress research in the UK (see §7.2), much of Vinogradov's research was contained in specialist military publications, which were inaccessible to the general public. However, even at that time, Vinogradov was conscious of the need to publish in International Journals, but making any headway in that direction involved overcoming grave official difficulties.

After the war, Vinogradov's interest turned to the rheology of polymers, but his first love for greases and contact friction and wear remained with him, and his last honour - the Gold Medal of the International Union of Tribology - was awarded to him in 1982 in recognition of his distinguished work in this field.

By the 60s, Vinogradov had risen to the rank of Colonel and it was only then that he was able to resign from the Armed Forces to join the Institute of Petrochemical Synthesis of the Academy of Sciences. He already had a contact with the Institute since he had been sent there a decade or so earlier to prepare his Doctor of Science dissertation[6].

Vinogradov's major shift of emphasis towards polymer rheology occurred in the late 50s, encouraged by an important Government resolution in 1957 about accelerating development of the Soviet chemical and polymer industries. This released funds to support the work and, in 1963, Vinogradov succeeded in setting up a Laboratory of Polymer Rheology, which he headed with distinction until his death on April 25th 1988, after enduring a long period of suffering with Parkinson's Disease.

The Laboratory of Polymer Rheology at the Institute of Petrochemical Synthesis was the first and, for many years, the only centre of rheological research in Russia, and Vinogradov was unquestionably the leading Russian rheologist of his day. This is reflected in the fact that the National Society has now been renamed the Vinogradov Rheology Society. At the same time, it is sad to note that the Soviet Union did not bestow many honours on Vinogradov and, apart from a military decoration, his only award was the Lebedev Honour Medal from the Academy of Sciences.

Vinogradov was interested in many aspects of rheology and his influential work on polymer rheology is covered in his well-received book on *"Rheology of polymers"* (English Translation by Mir Publishers Moscow, 1980), written with one of his most distinguished research students and subsequent collaborator, A Ya Malkin. Other well known rheologists who were taught by or associated with Vinogradov include A I Leonov and A I Isayev, both of whom later emigrated to the USA.

Vinogradov's wife, Kseniya Glebovan, was well known amongst the international rheology community as his constant companion, close friend and assistant. They had two daughters.

Photography was the favourite Vinogradov pastime and he was able to satisfy his desire to make colour slides of interesting places through his many international visits and activities. He belonged to that select group of Soviet scientists who were allowed to travel with relative freedom, although this must also be put into perspective, since his first overseas visit did not take place until he was nearly 50.

Georgii Vinogradov was a charming man. He was sociable and witty and he enjoyed the good things of life - although not, it seems, alcohol, after a surfeit during the war years.

---

[6]In the Soviet Union of Vinogradov's time and now in Russia, there were and are two higher degrees; a lower one roughly equivalent to a Western PhD and a higher DSc. Approximately one in ten with a PhD proceed to the second doctorate.

Chapter 7

Some Distinctive Rheological Concepts and Phenomena

7.1. Non–Dimensional Groups in Rheology

The utility of dimensionless groups in *Newtonian* fluid mechanics is well known, since the pioneering work of Reynolds (1883b). The governing equations for *elastic* liquids are far more complex than the Navier-Stokes equations and accordingly more non-dimensionless groups are required in rheology. At the simplest level, one might argue that the addition of a characteristic time λ as an extra parameter would, at most, add one more dimensionless group to the problem, but the very complexity of the equations ensures that non-dimensionalization is not as straightforward as in the Newtonian case, and only for the most restrictive of liquid types and flow regimes can the exercise be viewed as unambiguous (see, for example, Astarita 1979).

Notwithstanding these general difficulties, it is possible to highlight *two* rheological non-dimensionless numbers in particular, which have almost attained the level of prominence in rheology that the Reynolds number has been afforded in Newtonian fluid mechanics.

The first is called the Weissenberg number W_e. (Some writers, because of the prior use of W_e for the Weber number (ratio of surface tension forces to inertia) prefer to use W_i as the symbol.) Just as the Reynolds number is a convenient measure of the ratio of inertial to viscous forces, so the Weissenberg number can be viewed as a ratio of elastic to viscous effects.

The first exercise in using dimensional analysis in rheology is contained in a long paper by Herzog and Weissenberg (1928), in which they introduced a number representing the ratio of elastic to viscous effects (cf. White 1990). It was not until a generation later that Weissenberg returned to the subject (Weissenberg 1949) and emphasised the importance of 'recoverable shear strain', which for slow steady-shear flow is equivalent to the stress ratio $N_1/(2\sigma)$ (in the notation of Appendix 1).

In a study of the equations of *slow* elastico-viscous flow, White (1964) demonstrated the existence of a non-dimensional group, which we can write as $\lambda U/L$, where λ is a characteristic time of the liquid (expressible in terms of the linear-viscoelastic memory function), U is a characteristic velocity and L is a characteristic length. White called the group 'the Weissenberg number', since it has much in common with recoverable shear strain. This terminology has become firmly established in the rheological literature. The formal definition of W_e in terms of specific rheological functions has sometimes been modified, but the essential idea of W_e being the ratio of elastic to viscous effects has been maintained. Clearly, in all definitions, $W_e = 0$ for a Newtonian liquid.

The second important dimensionless number in rheology is called the Deborah number. Its origins can be traced unequivocally to Professor Marcus Reiner, who entertained a series of audiences in the late 50s and 60s with the background story. (Truesdell, 1980, recalls one such event at an after dinner speech at the 4th International Congress on rheology in Providence, RI, USA).

The basic gist of Reiner's talks has been published (Reiner 1964) and we can do no better than to quote from the published text: "In her famous song after the victory over the Philistines, Deborah sang 'The mountains flowed before the Lord'... Deborah knew two things. First that the mountains flow, as everything flows. But, secondly, that they flowed before the Lord, and not before man, for the simple reason that man in his short lifetime cannot see them flowing, while the time of observation of God is *infinite*. We may therefore well define as a nondimensional number the Deborah number

D_e = time of relaxation / time of observation.

The difference between solids and fluids is then defined by the magnitude of D_e. If your time of observation is very large, or, conversely, if the time of relaxation of the material under observation is very small, you see the material flowing. On the other hand, if the time of relaxation of the material is larger than your time of observation, the material, for all practical purposes, is a solid. In problems of industrial design, you may introduce the *time of service* for the time of observation. When designing a concrete bridge, you make up your mind to decide how long you expect it to serve, and then compare this time interval with the time of relaxation of concrete.

It therefore appears that the Deborah number is destined to become the fundamental number of rheology, bringing solids and fluids under a common concept... The greater the Deborah number, the more solid the material; the smaller the Deborah number the more fluid it is".

It is surprising that rheologists have been content with only two extra non-dimensional numbers. On occasions, the characteristic time of observation in the definition of the Deborah number has been taken as L/U, so that $D_e = \lambda U/L$; in these circumstances, the Weissenberg number and Deborah number have essentially the same definition and are used interchangeably, even though their origins are quite different.

7.2. The Weissenberg Effect

When a rod is rotated in a bath of Newtonian liquid, the free surface is depressed near the rotating rod due to inertial effects. In contrast, when a rod is rotated in a non-Newtonian liquid having measurable normal stresses, the free surface can rise spectacularly, with the liquid essentially climbing the rod. A full collection of pictures of the phenomenon is contained in the book by Boger and Walters (1993) and, in Figs 6.1 and 6.2 of the present book, we have included the classic pictures provided by Weissenberg (1947, 1949). The rod-climbing phenomenon is now generally referred to as the Weissenberg effect, although there has been some debate concerning the priority of discovery of the phenomenon (see, for example, Rivlin 1971). It is difficult to be precise about the matter, since the major events took place during the Second World War. However, a careful, hopefully unbiased, reading of the available literature and correspondence would lead one to the following conclusions:

(i) The original discovery arose from research work carried out during the war years on the saponified hydrocarbon gels used as flamethrower fuels.

(ii) The main contenders for priority of discovery would appear to be two laboratories in the UK. The first, headed by F H Garner, was located in the Chemical Engineering department of Birmingham University. Garner was assisted by A H Nissan, who was regarded as the rheology expert in the group. Also involved was one of Nissan's postgraduate students, G F Wood. Work on the flamethrower gels commenced in 1942 under a contract Garner held with the British Government. It continued until 1947. Not surprisingly much of the earlier work was classified as 'secret' and Garner, being meticulous about protocol, ensured that the research was not published until after the Second World War (see, for example, Garner, Nissan and Wood 1950). The second laboratory was headed by C H Lander at Imperial College, London (see Lander 1945). Work on the rod-climbing phenomenon took place there between 1943 and 1945 (see Weissenberg 1947). Weissenberg was an advisor to the group, although he had an appointment at the Shirley Institute, Manchester. The work was classified as secret, but monthly reports were issued and distributed amongst the various workers in the field in the UK (and the US) and the Birmingham and I.C. laboratories certainly knew of each other's research at the time (Weissenberg 1947, Nissan 1994).

(iii) Rivlin (1971) refers to a conversation he had with E K Rideal FRS, who had been in close contact with the I.C. programme in his official capacity as Advisor to the Ministry of Supply. Apparently, Rideal was under the impression that the rod-climbing effect had been discovered before Weissenberg's association with the group.

(iv) Nissan (1994) states that, irrespective of the general question of priority of discovery, the Birmingham group certainly discovered the rod-climbing effect for themselves. At the same time, Nissan hints that others, notably J Pryce-Jones, may have stumbled across the phenomenon earlier.

(v) Weissenberg was certainly very active in the immediate post war years in publicising the rod-climbing phenomenon, and it is for this reason, as much as any other, that the phenomenon has become known as the Weissenberg effect. Furthermore, there is general agreement that Weissenberg was the first to attribute rod-climbing to the existence of normal-stress differences (see, for example, Rivlin 1971, White 1990). So, for example, Weissenberg (1947) writes "If the lines of flow are closed circles, the pull along these lines strangulates the liquid and forces it inwards against the centrifugal forces and upwards against the forces of gravity. The presence and absence of such a strangulation is illustrated in Fig 1" (see our Fig 6.1).

(vi) The leading players in the unfolding story were apparently less interested in the question of 'priority' than some later workers in the field. A good example of this is provided by Nissan (1994) who, with the benefit of decades of hindsight, states "In the large-scale roll of the history of science, I do not think these 'priorities' are really significant".

(vii) The rod climbing phenomenon was certainly referred to as the Weissenberg effect as early as 1949 (cf. Reiner, Scott Blair and Hawley 1949) and this is now the accepted terminology. The Reiner et al (1949) paper also contained photographs of the effect in sweetened condensed milk and these are reproduced in Fig 7.1.

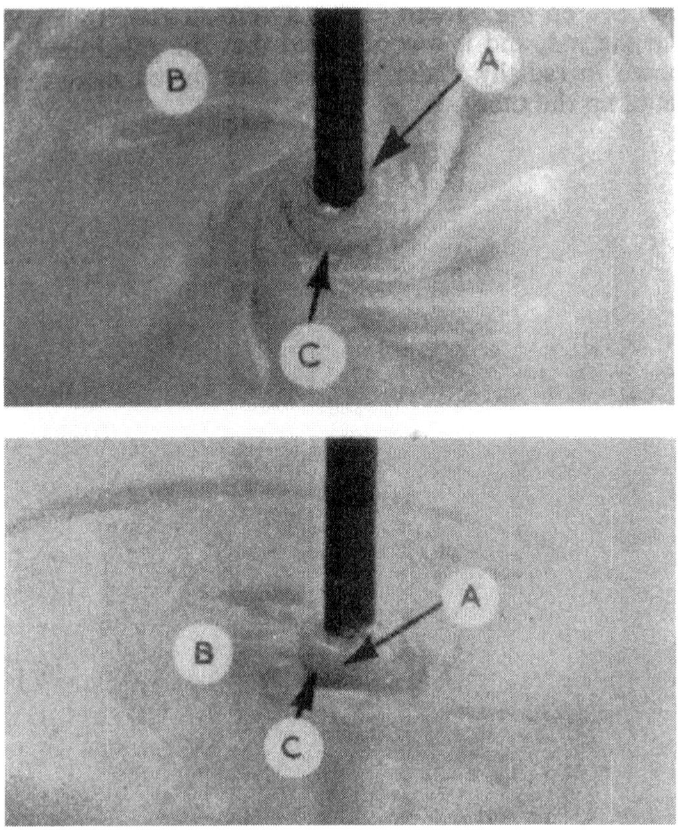

Fig 7.1. The Weissenberg Effect in sweetened condensed milk, shown from two angles. A rod has been rotated in the milk and the pictures taken after the movement was stopped following a few seconds rotation (Reiner et al 1949). The text says "A shows a globular mass of material drawn up the rod, B the lines along which the material is drawn to the centre, and C a crater formed by the materials being thrown outwards away from the rod by centrifugal forces".

The apparatus needed to produce the Weissenberg effect can be very simple indeed, since all that is essentially required is a beaker of test liquid and a rotating rod. Not surprisingly, therefore, the technique has become popular in illustrating the distinctive flow characteristics of non-Newtonian elastic liquids in a simple manner. At the same time, surprisingly little fundamental work has been carried out on the basic rod-climbing experiment, the notable exception being the careful experiments and supporting analysis of **Joseph** and his collaborators (see, for example, Joseph and Fosdick 1973, Joseph et al 1973, Beavers and Joseph 1975, 1977).

7.3. Extrudate Swell

It is well known to workers in classical Fluid Mechanics that, on exiting from a long capillary tube, a *Newtonian* liquid can either swell or contract. Specifically, if a swell ratio S_w is defined by $100(R-a)/\,a\,\%$, where R is the radius of the extruding liquid and a is the radius of the capillary, S_w can vary between approximately +13% for vanishingly small Reynolds numbers to approximately −13% for high Reynolds numbers (see, for example, Bird et al 1987a, p72).

It is equally well known to rheologists that, for highly elastic liquids, particularly polymeric liquids, S_w can take very high values indeed and swell ratios in excess of 200% are not uncommon. For this reason, extrudate swell is often used as a clear and dramatic demonstration of non-Newtonian behaviour.

There has been little controversy concerning the *possibility* of very large extrudate swell for highly elastic liquids, but correlating this unambiguously with well defined rheometrical functions is difficult. Certainly, the normal-stress differences N_1 and N_2 are important, but other factors, such as the length of the capillary tube, viscous heating and general memory effects can also play a role (see, for example, Tanner 1970, 1988, Bird et al 1987a).

The present book makes it abundantly clear that assigning priority of discovery to rheological phenomena is often controversial and 'extrudate swell' is no exception. Some have used the term 'Barus effect', whilst others have argued strongly in favour of a much later discovery, assigning the term 'Merrington effect' to the phenomenon. An example of this controversy is contained in correspondence in the Bulletin of the British Society of Rheology.

In 1968, Rigby objected to the term 'Barus effect', arguing instead in favour of 'Merrington effect', supporting a proposition to that effect by Truesdell and Noll (1965, p458). Rigby's stance was strongly contested by MacSporran (1969) in the next issue of the BSR Bulletin. He concluded that Merrington was by no means the first to observe the phenomenon and wrote that "a more careful search of the literature reveals that the post-extrusion swelling effect was well known long before Merrington's work was published. It is also apparent from the literature that the phenomenon reported by Barus is the same as that observed by the later workers, including Merrington".

To put such comments into perspective, it is instructive to summarize the work of both Barus and Merrington.

Barus (1893) conducted experiments on marine glue and his published works contain the pertinent comments: "At the end of stated intervals of time (usually hours), the small cylinders of marine glue which had exuded were cut off with a sharp knife and weighed. Now it was curious to note that these cylinders, left to themselves for about a day, showed a gradual and marked deformation, such that the originally plane bottom or surface of section eventually expanded into a symmetrical projecting conoid, with an acute apex angle of less than 45°. I take this to be an example of volume viscosity. Thus the experiment points out in a beautiful way how much residual viscosity resembles a slowly reacting elasticity".

Much later, Merrington (1943) conducted experiments on rubber solutions and also mineral oils containing metallic soaps. His short 'Nature' article contains a figure (Fig 7.2) which gives a clear picture of extrudate swell. The supporting text includes a possible explanation of the phenomenon: "When the liquid emerges from the capillary, it is still elastically deformed and will consequently tend to return to its former 'unstretched' condition. This results in a swelling of the emergent column of liquid which is easily observed, as Fig 2 shows[4]. It is evident that elastic recovery commences at the actual capillary end... it is to be expected that this swelling will increase both with increasing stress and increasing concentration, and this is found to be so for the materials in question.

... the swelling of the emergent column of liquid ... should give a measure of the elastic property of the material". Note that the reference[4] in Merrington's text is given as:

[4]Barus phenomenon. *Amer J Sci 45*, 87 (1893).

The following observations are in order at this point:

1. The Barus effect *may* be related to extrudate swell as currently understood, but it is clearly a different phenomenon, possibly recovery of shear.

2. Merrington has confused the picture himself by *implying* that the phenomenon he observed was the same as that presented by Barus.

3. Extrudate swell seems to have been taken for granted by some researchers in the inter-war years. So, for example, Halton and Scott Blair (1936) report: 'During extrusion, the dough cylinder swells, this swelling being, in general, greater for good than for poor quality doughs'. A year later, Houwink (1937) noted: "In practice the elastic phenomena during extrusion play an important role. In every rubber factory it is well known that during tubing the extruded rubber has a longer diameter than that of the nozzle". Later still, Garner, Nissan and Wood (1950, p38-40) noted that they had independently observed the swelling effect in the 1942-46 period, but did not publish due to the constraints of the war years.

4, Merrington's Nature article contains the first recorded picture of the phenomenon of extrudate swell as now understood, i.e. the isothermal swelling of an extrudate on emerging from a long capillary tube.

5. The 'terminology' component of the controversy has been somewhat overtaken by events, since both 'Barus effect' and 'Merrington effect' are now out of vogue and 'extrudate swell' or 'die swell' are the most common current expressions for the phenomenon.

6. Merrington (1943) was careful to point out that the swelling in *his* experiments commenced at 'the actual capillary end'. Some time later, Giesekus (1968) gave pictorial evidence that, when the Reynolds number is high enough, the swelling can be delayed (see Fig 7.3).

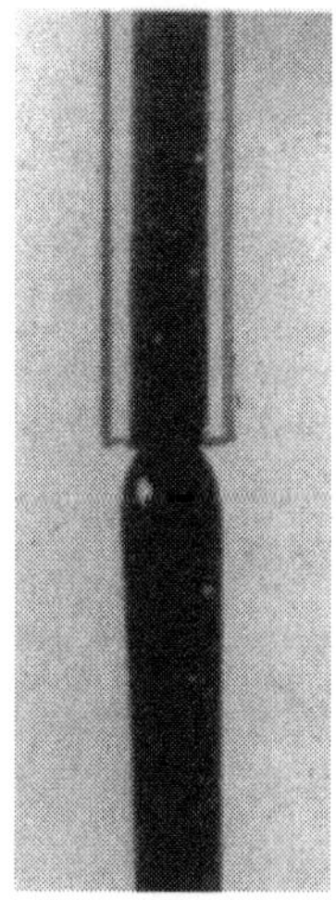

Fig 7.2. Merrington's well known photograph of extrudate swell. (From Nature **152**, 1943, 663.)

7.4. The Tubeless Syphon

We have already indicated in Chapter 6, that for some dilute polymer solutions, the *extensional* viscosity η_E can be very high indeed, with the associated Trouton ratios being greatly in excess of the 3 expected of Newtonian liquids. This feature of non-Newtonian behaviour manifests itself in a variety of ways, some of which can be quite spectacular. The more photogenic of the phenomena have been captured in the book of Boger and Walters (1993).

The so-called 'tubeless syphon' is a particularly impressive experiment, which is regularly used to demonstrate excessive resistance to an extensional flow in some elastic liquids. Here, fluid from a reservoir is drawn up through a nozzle by a vacuum pump and the nozzle is then raised above the level of the liquid in the reservoir. For many elastic liquids, the upward flow continues, even when the nozzle is several centimetres above the free surface.

According to Bird et al (1987a) the phenomenon was used in connection with the harvesting of bitumen on the Dead Sea around AD70, the relevant reference being to *the Complete Works of Tacitus*, who was born in AD55. In more recent times, the phenomenon was discussed by Fano (1908) and it is sometimes called 'Fano flow'.

In his paper, Fano perceptively noted that "spinnable materials are distinguished above all by the fact that, under suitable conditions, they show elasticity in tension to such an

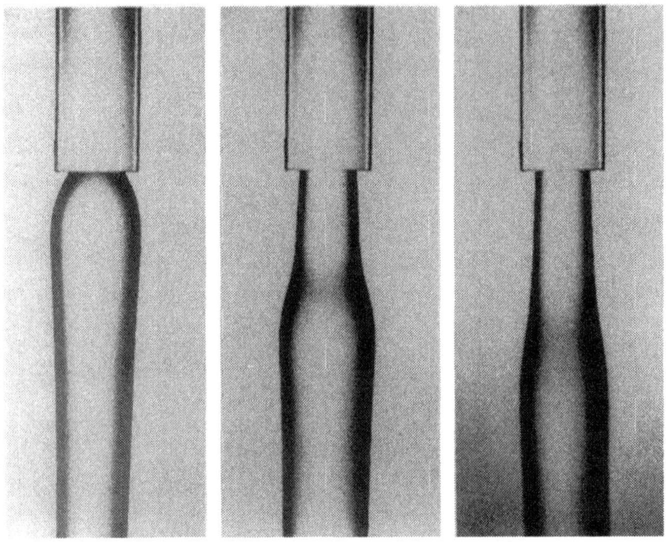

Fig 7.3. Delayed die swell for a 5% aqueous solution of polyacrylamide. (a) Normal die swell. (b) Moderate flow strength. (c) Strong flow. (From H Giesekus, Rheologica Acta **8**, 1968, 411-421.)

extent that one is led to consider them, in respect of some properties at least,to be more like solids than liquids". This anticipated Reiner's 'Deborah number' ideas by several decades, although Reiner would no doubt have appealed to the Old Testament for precedence.

An interesting variant of Fano flow was described by James in 1966. This involved the syphoning of a liquid from one beaker to a lower one, the liquid running up the side of the upper beaker before falling freely into the lower one (see Fig 7.4). James referred to the flow as an 'open channel syphon'. It has become a popular and dramatic demonstration of fluid elasticity.

7.5. Thixotropy

'Thixotropy' is a distinctively rheological phenomenon, which is found in many complex materials, especially colloidal systems. It is associated with changes in rheological properties with time, even when the applied stress or strain rate is kept constant. In that sense, it is quite different from 'shear thinning', although it is easy to appreciate how the two can be and have been confused, the confusion being compounded by the fact that the same material can exhibit both thixotropy and shear thinning. Unfortunately, the confusion has crept into the rheological literature. So, for example, Chambers Dictionary of Science and Technology (1971) defines Thixotropy as a "rheological property of fluids and plastic solids characterized by a high viscosity at low stress, but a decreased viscosity when an increased stress is applied". This is, of course, a standard definition of 'shear

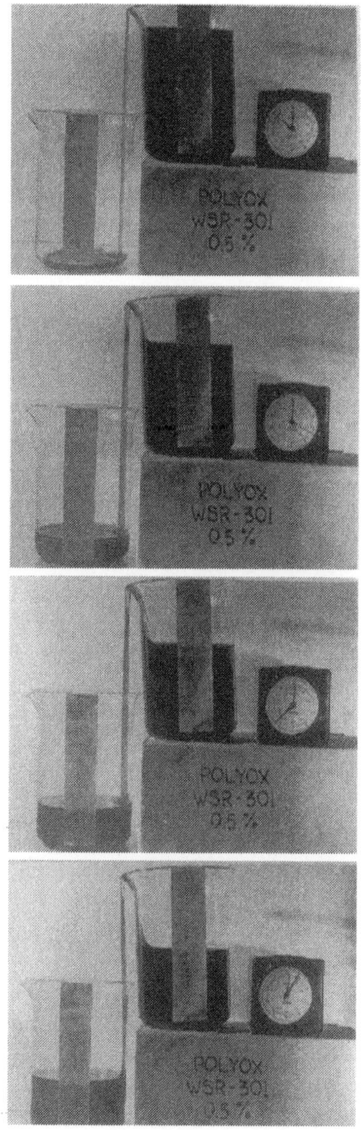

Fig 7.4. The open channel syphon. The viscoelastic fluid is a 0.5% (by weight) solution of polyethylene oxide in water, the polymer having a mean molecular weight of 3.8×10^6. The viscosity of the liquid at 23°C is 80 centipoise. The scale on the 4 litre beakers is in inches, and one revolution on the clock is 10 sec, the entire sequence shown taking 13 sec. A dye has been added to the fluid to show the effect more clearly. (From D F James, Nature **212**, 1966, 754-756.)

thinning' and has nothing to do with thixotropy. A more recent definition is closer to the mark (Barnes et al 1989) "A decrease of apparent viscosity under constant shear stress or shear rate, followed by a gradual recovery when the stress or shear rate is removed. The effect is time dependent". In the sense that many natural materials exhibit thixotropy, the phenomenon itself must have been known centuries ago. However, Barnes (1997), in a comprehensive study of thixotropy, points to the work of Schalek and Szegvari (1923) as heralding modern interest in the subject. These authors observed that aqueous iron oxide gels "have the remarkable property of becoming completely liquid through gentle shaking alone". A little later the term thixotropy was coined by Peterfi (1927) on the basis of the Greek words *thixis* (stirring or shaking) and *trepo* (turning or changing). In 1935, Freundlich published a book entitled *"Thixotropie"*. Scott Blair, whose research ranged over many materials that could be considered thixotropic, commented as late as 1943 that the whole subject of thixotropy was 'so very new'. However, Barnes (1997) perceptively points out that Scott Blair nevertheless listed over 80 papers on the subject. In the same book, Scott Blair (1943) provided a list of papers on the so-called 'thixotrometers', instruments to characterize the phenomenon. This terminology is now virtually obsolete, unlike the term thixotropy itself, which is one of the few pre-Second World War rheological expressions to have survived. The modern generally agreed definitions of shear thinning and thixotropy are unambiguous and distinct and in that sense no confusion should now remain. However, as the field has evolved, numerous researchers have, with some cause, viewed the two phenomena as being interrelated. So, for example, Scott Blair (1943) comments that "With our present knowledge of microstructural changes, it is probably safe to say that all materials that are shear thinning are thixotropic, in that they will always take a finite time to bring about the rearrangements needed in the microstructural elements that result in shear thinning. If this recovery is very rapid, the phenomenon is observed as [shear thinning]; if slow, it is observed as thixotropy". In the same vein, Jobling and Roberts (1957) wrote that "thixotropy now has an even less distinct connotation. Electronic methods of measurement have shown that the time-lag required before the original structure is regained may be very short indeed and it then becomes difficult to distinguish between a thixotropic material with a very short recovery time and a material whose viscosity falls with increasing rate of shear and depends for all practical purposes only on the instantaneous rate of shear". In the Discussion to the Jobling and Roberts paper, Reiner remarked that [shear thinning] and thixotropy are seen by some as being the same thing, with shear thinning seen as a material with 'nearly zero time of recovery'.

It is now well known that some time-dependent materials show the opposite type of response, with the viscosity under a constant stress or strain rate *increasing* with time of shearing, with a gradual recovery when the shearing is stopped. A graphic example of the phenomenon is shown in Fig 7.5, taken from an expanded version of a short publication by Cheng (1973).

The phenomenon is known by many names and the terminology has been confusing to say the least. The use of 'rheopexy' in this connection has been deprecated by Reiner and Scott Blair (1967), who point out that the definition of rheopexy by Juliusburger and Pirquet (1936) refers to 'the solidification of *thixotropic* sols by gentle and regular movement'. Accordingly, Reiner and Scott Blair argue that it should not be used for non-thixotropic systems, a view supported by Cheng and Evans (1965).

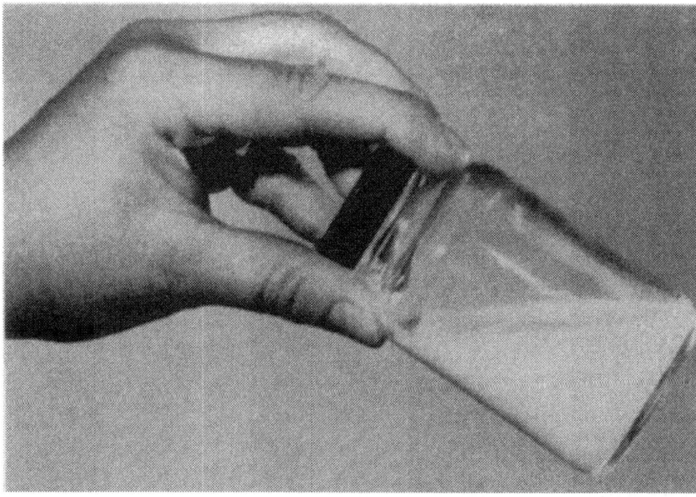

a) AFTER PROLONGED REST

b) IMMEDIATELY AFTER VIGOROUS SHAKING

Fig 7.5. An example of antithixotropy. The liquid is an alkaline perbunan latex (Cheng 1973).

Another term to complicate the terminology is 'dilatancy'. The term was originally coined by Osborne Reynolds (1885) in connection with the observation that, when walking on moist sand on the beach, the sand under the foot becomes firm and hard, becoming moist again when the pressure due to the foot is removed. A volume expansion is clearly involved. The original definition of dilatancy was extended in the 1930s to include any system which increases in viscosity with increasing rate of shear;[1] it was further broadened (unhelpfully) to represent the opposite of thixotropy (see, for example, the discussion in Bauer and Collins, 1967).

On a more constructive note, Cheng and Evans (1965) include a short discussion on the terminology of time-dependent materials. They point out that the opposite of thixotropy was termed 'negative thixotropy' by Hartley (1938) and 'antithixotropy' by l'Hermite (1949), both terms implying that 'the structure builds up under shear but breaks down at rest'. Either term is clearly appropriate to describe the phenomenon, although Cheng and Evans appear to favour 'antithixotropy'.

7.6. Instability in Flow

7.6.1. Introduction

Newtonian flows often display instability at a critical flow speed or, more strictly, at a critical Reynolds number. Perhaps the most important case is the transition from Poiseuille flow to turbulent flow in pipes and channels, at a Reynolds number of around 2000. Characteristically, the simple parabolic Poiseuille flow changes to a much more complex chaotic time-dependent flow at the critical speed. Sometimes, for example in Couette flow between a rotating inner cylinder and a fixed outer cylinder, transition occurs to another steady, non-chaotic flow.

The above two examples show the interplay between viscous and inertia forces. Other instabilities due to surface tension, free surfaces, and gravity occur in Newtonian fluid mechanics and are detailed in several books (see for example, Chandrasekhar 1961, Drazin and Reid 1981 and Joseph 1976a, b).

For non-Newtonian flows, the extra dimensions of shear-thinning, elasticity and normal-stress effects all offer additional possibilities for unstable behaviour and the purpose of this section is to describe progress in this area. The search for the critical point at which a change from a simple flow to a more complex one occurs involves the solution of very difficult eigenvalue problems, even for quite simple cases, and we shall not discuss these in any great detail. We will term the simple flow the 'base flow' and analysis then seeks a perturbation of this base flow.

For example, if the base flow field is denoted by $\boldsymbol{V}(\boldsymbol{x})$, and the perturbed flow by $\epsilon\boldsymbol{v}(\boldsymbol{x}, t)$, where in linear stability analysis $\epsilon << 1$, emphasis is often focused on perturbations of the form $\boldsymbol{v} = e^{st}\boldsymbol{\omega}(\boldsymbol{x})$. The eigenvalue problem then seeks to determine conditions for:

(i) The real part of s to be positive (corresponding to the unstable case), negative (corresponding to stability) or zero for so called 'neutral stability'.

[1] A comprehensive review of 'shear thickening ("dilatancy") in suspensions' is given in Barnes (1989).

(ii) The imaginary part of s is zero or non-zero. In the latter case, oscillation sets in at the critical condition and the secondary flow is unsteady. This is often called 'overstability'.

If the flow speed is such that the critical conditions are exceeded, the secondary flow will grow and may lead to even more complex behaviour as the speed of flow is increased further. Thus considerable interest lies in finding the point of and manner of bifurcation from the original simple flow. Details of many examples of viscoelastic instabilities have been published by Larson[2] (1992) (see also Tanner 1988).

In the non-Newtonian case, two sets of artefacts not present in the Newtonian case have appeared in the literature, and these should not be confused with the type of genuine physical instability just described. In the first case, some numerical instabilities have arisen and practitioners have been tempted to suggest that these are part of the real flow system, but in most cases they are resolvable by more careful numerical work. In the second case, the possibility arises that the constitutive equation itself contains features that lead to instability. A famous example of this is Joseph's (1981) analysis showing that all n^{th} order fluids (§4.4) are intrinsically unstable; the use of these models in analyses needs to be very carefully circumscribed and confined to very slow and very slowly varying flows. Another example concerns models in which the shear-stress/shear-rate curve reaches a maximum. Such models will demonstrate instability even in a perturbed simple shearing flow (Tanner 1988) and they have been introduced deliberately into some analyses to simulate 'spurt' flow (see Larson 1992). We will not discuss these two cases here, even though they are of interest in numerical work.

With these preliminary remarks, we will now display some selected results; a study of Larson (1992), Shaqfeh (1996) and Huilgol and Phan-Thien (1997) will reveal many more cases of interest.

7.6.2. Some Early Results

The 1960's saw an upsurge of interest in viscoelastic instabilities. We believe the first papers to consider stability in nonlinear viscoelastic fluid models were the series published by Walters and co-workers (Walters 1962, Thomas and Walters 1963, 1964a, b, Chan Man Fong and Walters 1965, Chan Man Fong 1965, Beard et al 1966), in which the so called Dean problem, the Taylor-Couette problem, and the analogue of the Orr-Sommerfeld equation for elastic liquids of the Oldroyd-A and Oldroyd-B types were investigated. In the work of Beard et al, it was shown that, under some circumstances, instability could set in as an oscillatory-in-time mode corresponding to overstability; this is quite different to the *steady* secondary flow predicted in the Newtonian case under comparable conditions.

Datta (1964) and Rao (1964) discussed the second-order fluid model and showed that the second normal stress difference N_2 is a dominant factor in Couette stability. Ginn and

[2]Ronald Gary Larson was born in Litchfield, Minnesota in March 1953. His university education was confined to the University of Minnesota, where he obtained a BS in 1975, an MS in 1977, and a PhD in 1980, all in Chemical Engineering. As a graduate student, he was supervised by L E Scriven and H T Davis. In 1980, he joined the staff at Bell Laboratories, Murray Hill and he was promoted to the status of 'Distinguished Member of the Technical Staff at AT and T Bell Laboratories' in 1987. He resigned in October 1996 to take up an appointment as Professor of Chemical Engineering at the University of Michigan, Ann Arbor.

Denn (1969) used the CEF model (eq. 4.7) and showed that a negative N_2 was stabilizing, but that a positive N_1 was destabilizing. Miller and Goddard (1967) had carried out an impressive analysis using a complex co-rotational model, but this was not published until 1979 (Goddard 1979), by which time Karlsson et al (1971) had used the KBKZ model to show that the choice of model was not exceptionally critical if the instability took the form of a steady secondary flow. There were (and are) some problems in reconciling the cell spacing observed in experiments with existing analyses (Tanner 1988) and these have often been ascribed to finite-length effects, not modelled in the analyses, or to nonlinear effects.

Earlier, in 1966, Giesekus had studied the Taylor-Couette flow experimentally at very small Reynolds numbers, so that the inertia-driven instabilities discussed above did not occur. He concluded that there might be a purely elastic instability, and Larson et al (1990) later predicted such a result using an Oldroyd-B model. Avgousti and Beris (1993) have obtained numerical results for this case and the results are quite complex; a non-axisymmetric mode can be the least stable. Larson (1992) also describes some relevant experiments.

One of the simplest flows to analyse is the Bénard heat conduction problem, where the base flow is a state of rest. Aside from a distant possibility of overstability (Vest and Arpaci 1969, Sokolov and Tanner 1972), the problem is chiefly of interest for the generality of the solutions possible.

The plane Poiseuille flow problem for elastic liquids has proved to be a difficult one, but Ho and Denn (1977/78) did obtain a solution for the UCM model. Their numerical analysis had to be fairly elaborate, but Renardy (1992) confirmed this work analytically.

7.6.3. Cone-and-Plate and Torsional Flows

Experimental rheologists have an obvious interest in cone-and-plate and torsional flows, since these are regularly employed to yield viscometric data (see Chapter 6). Not surprisingly, therefore, the stability or otherwise of such flows is a major concern. One important phenomenon, which can be viewed as a species of instability, is called 'shear fracture'. Here, the edge of the sample in the rheometer is severely distorted, causing at worst an ejection of the sample at quite low Weissenberg numbers and at best a point beyond which measurements cannot be believed. An early attempt to explain the phenomenon was made by Hutton (1965) on the basis of elastic energy associated with the first normal stress difference N_1. Tanner and Keentok (1983) put forward an alternative explanation starting from the idea of a semi-circular 'flaw' at the edge. It turned out that the balance between N_2 and surface tension was crucial here, with fracture occuring when

$$-N_2 > \frac{2\Gamma}{3a}, \tag{7.1}$$

where Γ is the surface tension coefficient and a is the radius of the semi-circular flaw. Experiments by Lee et al (1992b) appear to confirm (7.1).

Cone-and-plate and torsional flows have also given rise to other unusual instabilities. Jackson et al (1984) observed what appeared to be an unexpected antithixotropic shear-thickening in experiments on Boger fluids. Up to that time, the attraction of Boger fluids had been their almost constant shear viscosity. The group observed that, at high shear rates, measurements of the torque and the normal force in a Weissenberg Rheogoniometer increased steadily at a fixed shear rate over a long period of time; the data for torsional flow (reproduced in Fig 7.6) showed nearly double the torque expected for the final long-term 'equilibrium' values. Jackson et al noted that halving the rheometer gap, keeping the maximum shear rate constant, removed the 'antithixotropy'.

At the same time, Phan-Thien (1983, 1985) carried out a stability analysis for the Oldroyd-B fluid in torsional and cone-and-plate flows and predicted the onset of instability (in inertialess flow) at a critical Deborah number. For torsional flow between infinite plates, he obtained

$$D_{e(crit)} = \pi \left[(1-\beta)(5-2\beta)\right]^{-1/2}, \tag{7.2}$$

where $D_e = \lambda\Omega$, Ω being the rate of rotation in this case, and β is the ratio of solvent viscosity to total viscosity. A similar formula was given for the cone-and-plate case (Phan-Thien 1985). Research by Magda and Larson (1988) and McKinley et al (1991) showed that the antithixotropy was a manifestation of an instability, similar to that predicted by Phan-Thien. However, the flow patterns did not resemble those of the analysis and so they concluded that the agreement between the predicted and observed Deborah numbers was fortuitous. As a further twist, Avagliano and Phan-Thien (1996) were later moved to analyse the stability of torsional flow between plates of *finite radius*, and this has given more satisfactory agreement with experiment.

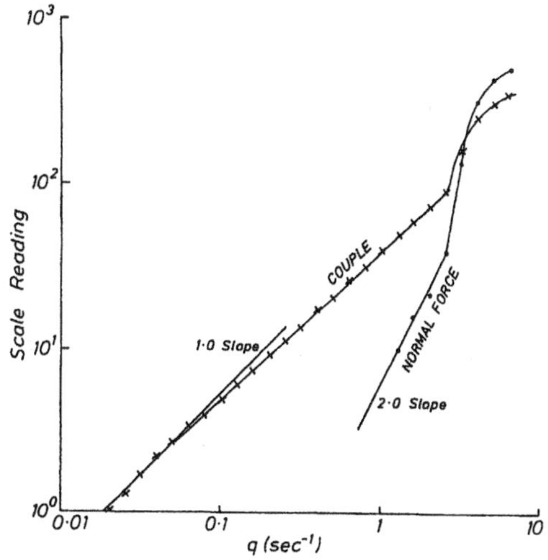

Fig 7.6. Torsional-flow data for a Boger fluid obtained from a Weissenberg Rheogoniometer. Here, q is the shear rate (from Jackson et al 1984); note the increase in torque and normal force at a critical shear rate of about $2s^{-1}$.

7.6.4. Extrudate Distortion and Fracture

We have already discussed one rheological phenomenon associated with the extrusion of elastic liquids from slits and capillaries, namely die-swell (§7.3). This is, however, superseded as the flow rate is increased by a succession of unstable flows, which give extrusion flows a particular richness and variety, something which has been of lasting interest to academic rheologists and industrialists alike (see, for example, Petrie and Denn 1976).

The first variation from a smooth extrudate manifests itself as a 'superficial scratching', or 'matting'. A 'striping' along the flow direction is also possible (see, for example, Benbow et al 1960, Giesekus 1972).

Later, the phenomenon known as 'sharkskin' can occur in which the extrudate becomes rough, with a small amplitude and high frequency disturbance. This is sometimes followed by a 'stick-slip' or 'spurt' flow (cf. Vinogradov et al 1972, Denn 1990), and as the flow rate is further increased, the dramatic instability known as 'melt fracture' can occur, usually accompanied by a helical screw-like extrudate of varying pitch and amplitude with gross distortions (cf. White 1990). This is often accompanied by an audible report (cf. Cogswell 1981). Initially, the term 'elastic turbulence' was used for the phenomenon, but this terminology has fallen out of favour.

The various instabilities mentioned above may not all be found as the flow rate is progressively increased for one particular polymer. For example, sharkskin is frequently absent in experiments and only linear polymers seem to exhibit a stick-slip region (Denn 1990).

It would appear that distorted extrudates were first reported by Nason (1945), but the leading players in the unfolding story were Spencer and Dillon (1949), Tordella (1956, 1957, 1958, 1963) and Benbow (see, for example, Benbow et al 1960, Howells and Benbow 1962, Benbow and Lamb 1963). In more recent times, M M Denn of Berkeley and J-M Piau[3] of Grenoble have devoted a considerable effort to the general problem and have published extensively on the subject (see, for example, Kalika and Denn 1987, Denn 1990, Piau et al 1990, El Kissi and Piau 1996).

The extrusion instabilities have always been amongst the most photogenic of rheological phenomena and we show in Fig 7.7 some early photographs of Benbow et al (1960) for two (linear) silicone gums at room temperature. In these, we see examples of sharkskin and melt fracture.

Controversy still exists regarding the site of initiation and the physical mechanisms of the various instabilities. In a relatively recent review article, Larson (1992) points to the general agreement that sharkskin is initiated at or near the die exit, and that melt fracture, when it occurs, affects not only the appearance of the extrudate but also the flow in the capillary die and in the die entry region (see also Piau et al 1990).

The mechanisms of gross melt fracture is a subject of ongoing interest to rheologists.

7.6.5. Instabilities in Extensional Flows

Instabilities can occur in *shear-free* flows, such as those occurring in fibre-drawing and film-blowing. Considering the case of fibre-spinning, we may note the reviews of Petrie

[3]Jean-Michel Piau was born in Doué la Fontaine, Maine et Loire, France on 18th October 1942. He obtained his doctorate from the University of Paris VI in 1973. From 1973 to 1976, he lectured at the IUT, Orleans, before moving to l'Institut National Polytechnique de Grenoble as Professor and Director of the Rheology Laboratory. He is currently Vice President of the European Society of Rheology. Professor Piau is married to Dr Monique Piau who is also a well known rheologist.

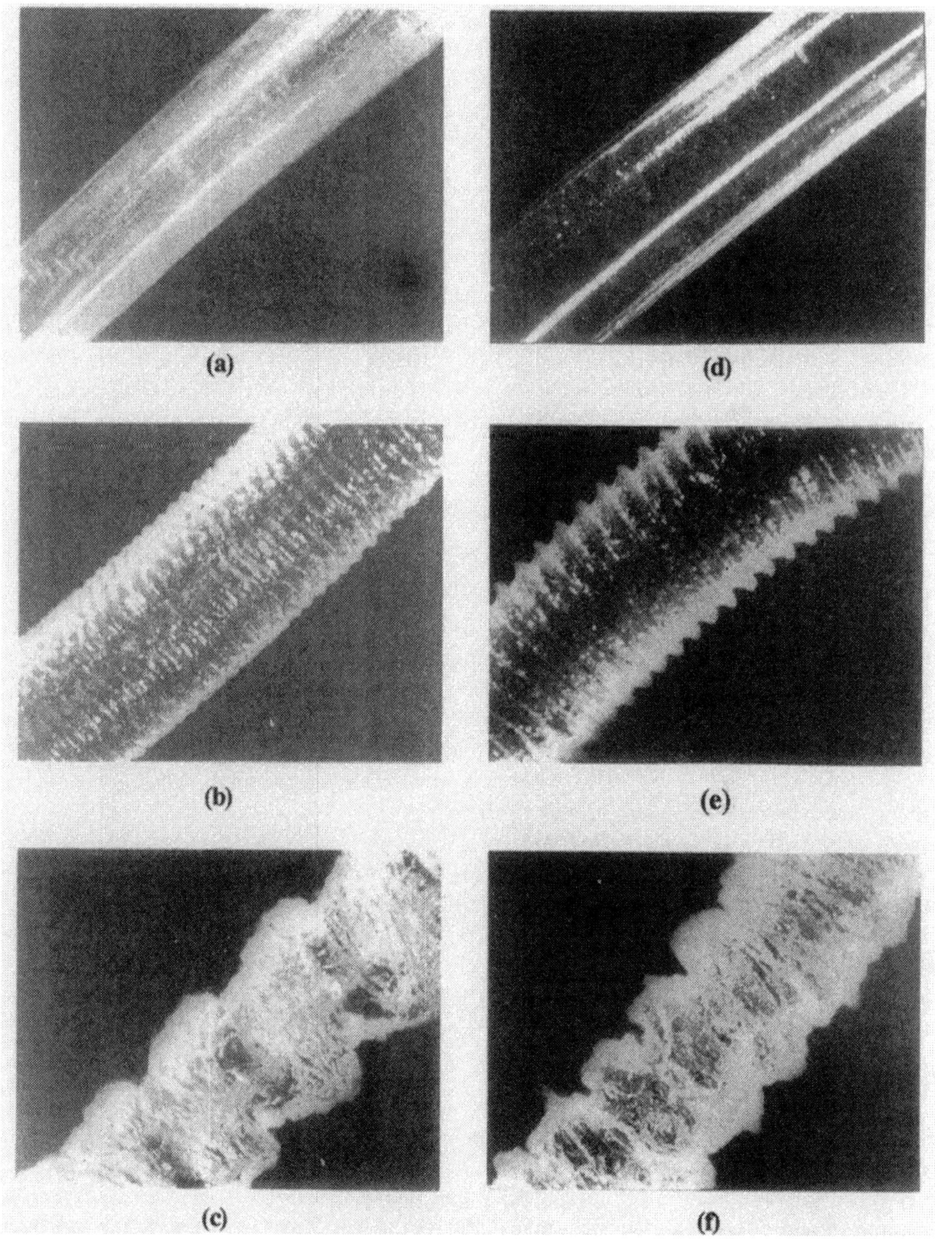

Fig 7.7. The extrudates with increasing flow rate for two silicone gums. (a), (b) and (c) are for a pure dimethyl siloxane. (d), (e) and (f) are for a PDMS (with a percentage of phenyl groups). (a) and (d) are called 'smooth', (b) and (e) sharkskin, (c) sharkskin plus melt fracture, (f) sharkskin plus some melt fracture. (From Benbow et al 1960.)

and Denn (1976) and Denn (1980). In the Newtonian case, one encounters the phenomenon of draw resonance, where small fluctuations in fibre diameter are amplified into a sinusoidal pattern along the length of the fibre. Kase et al (1966) and Matovich and Pearson (1969) analysed this phenomenon, neglecting inertia, air drag, gravity and surface tension. Stability analyses for viscoelastic constitutive equations have been carried out by Fisher and Denn (1975, 1976) and Chang and Denn (1980). Results have been obtained for the UCM fluid, and the curious feature here is that the viscoelastic effects are stabilizing, which is in contrast to many of the results reported above. A similar result is obtained when considering the viscoelastic analogue of jet break-up. Rayleigh (1879) showed that a cylinder of (inviscid) fluid with a surface tension breaks up into drops as instability grows. In the viscoelastic case, drops often continue to be connected to one another by extending filaments (Bousfield et al 1986), so that break-up is much reduced.

In summary, one sees that instability may be promoted or retarded by viscoelastic effects, depending on the base flow. It is also clear that many effects still defy rational analysis.

7.7. Drag Reduction in Turbulent Flow

Turbulent drag reduction is defined as 'the increase in pumpability of a fluid caused by the addition of small amounts of another substance to the fluid. During drag reduction, therefore, the fluid with additive requires a lower pressure gradient to move it at a given bulk mean velocity in a pipe than the same fluid without additive' (Savins 1964; see also Patterson et al 1969). The effect can be quite dramatic and, given the right conditions, reductions in drag of 70% are easily attainable. The phenomenon is usually referred to as 'the Toms effect' after the first scientist to clearly recognize and describe the phenomenon in the open literature (Toms 1949), although, not for the first time in rheology, there has been some debate concerning 'priority of discovery'.

Before the seminal work of Reynolds in 1883, there was no clear recognition of the existence of turbulent, as opposed to laminar, flow. Stokes (1845) was therefore not able to reconcile his laminar viscous tube-flow theory with experiments (see §1.3). Reynolds (1883a,b) showed the existence of the two regimes and studied the transition from one flow pattern to the other. He showed that transition to turbulent flow took place when the ratio of inertia forces to viscous forces reached a value of about 2000 in pipe flow; this ratio is now called the Reynolds number (R_e) and is defined for tube flow by the rule

$$R_e = \rho \bar{u} d / \eta, \tag{7.3}$$

where ρ is the fluid density, $\bar{u}$ is the average speed in the tube, d is the diameter, and η is the (Newtonian) viscosity.

Beyond this transition point, a stability limit, turbulence rapidly develops and the flow becomes chaotic. Each particle therefore undergoes a rich kinematic experience. We shall be mostly concerned with pipe and channel flows here, since they form the largest set of drag-reduction data.

Turbulent flows are irregular, diffusive, three-dimensional and dissipative. No complete analytical theory exists, but a great deal is known about such flows. A simple introduction to the subject of turbulence has been given by Tennekes and Lumley (1972).

We show in Fig 7.8 an example of drag reduction, taken from the first publication of the phenomenon involving macromolecules (Toms 1949). It shows the increase in flow rate, at small concentrations of polymer, for a fixed pressure gradient in the turbulent regime. This paper clearly recognized that the *turbulent* flow was affected by the addition of small amounts of macromolecules. However, on the matter of priority of discovery, we should also mention an early brief paper published in 1931 by Forrest and Grierson on the reduction of energy needed to pump wood pulp, which also seems to be a genuine example of drag reduction. (See Radin et al 1975).

Independent discoveries of the drag-reduction phenomenon in polymer solutions seems to have been made in both the United Kingdom and the United States during the 1939-1945 war, in the course of studies of the flow of thickened petroleum products (napalm) (cf. §6.1 and §7.2). Dr A H Nissan, in a letter to Dr Ralph Oliver of Birmingham University dated 18 March 1993, states that he worked in Professor Garner's group at the time (1942-1947) they were experimenting with petroleum fluids at Birmingham University and that he personally knew of drag reduction, because of design calculations that the group were making during the war. The work was written up in an internal report, but it was never cleared for publication.

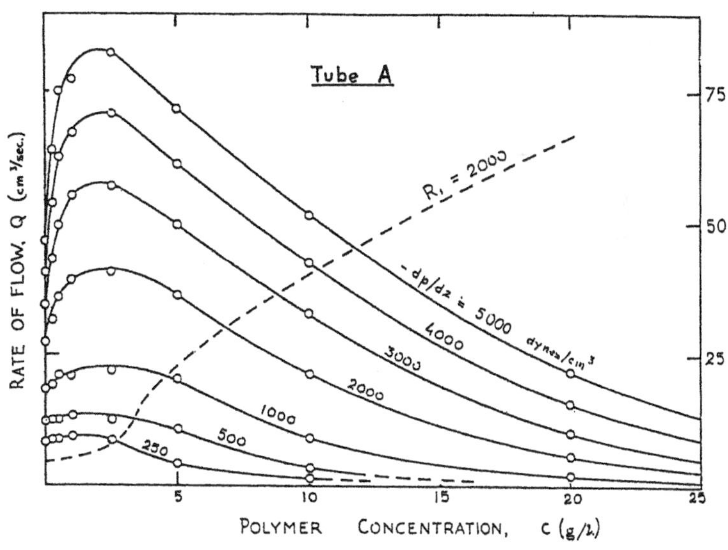

Fig 7.8. Rate of flow versus polymer concentration at different pressure gradients in capillary flow for solutions of polymethyl methacrylate in monochlorbenzene. (From Toms 1949.)

Similarly, Karol Mysels (1971) relates his experiences in this area. Mysels worked at the Edgewood Arsenal in Maryland during the war. In mid-1945, the work involved checking some surprising results obtained with a gelled gasoline (napalm). W A Klemm and G

A Agoston appear to have made the observations under the direction of Mysels. The report was released for public viewing to the Library of Congress on February 2nd, 1945. Subsequently, a US patent was granted to Mysels and a paper was published 9 years later (Agoston et al 1954).

Agoston et al (1954) tested the pressure/flow relations of 100g/ℓ and 20g/ℓ gels in a 1/8 inch diameter (3.2mm) pipe. The results are contained in Fig 7.9, which shows the lower pressure needed to send the gel through the tube compared to neat gasoline. The authors speak of the phenomenon as 'the apparent reduction of viscosity by the addition of a thickening agent'. The $20g/\ell$ results in Fig 7.9(b) seem to indicate a turbulent flow, but the $100g/\ell$ results in Fig 7.9(a) do not unequivocally indicate that the flow was turbulent; they appear to suggest a shear-thinning laminar flow. It is not therefore entirely clear that fully-developed turbulence existed in this set of experiments and we may well be seeing a laminar flow or a delayed transition to turbulence (cf. Lumley 1969); certainly pressure drops less than the turbulent neat-gasoline case are seen, but the pressure drop is considerably in excess (by about 50 times) of the pressure loss that would exist in a laminar flow of gasoline (the dotted line in 7.9). The authors remark that 'the flow of gasoline is turbulent, but no information is available about the flow of the jelly'.

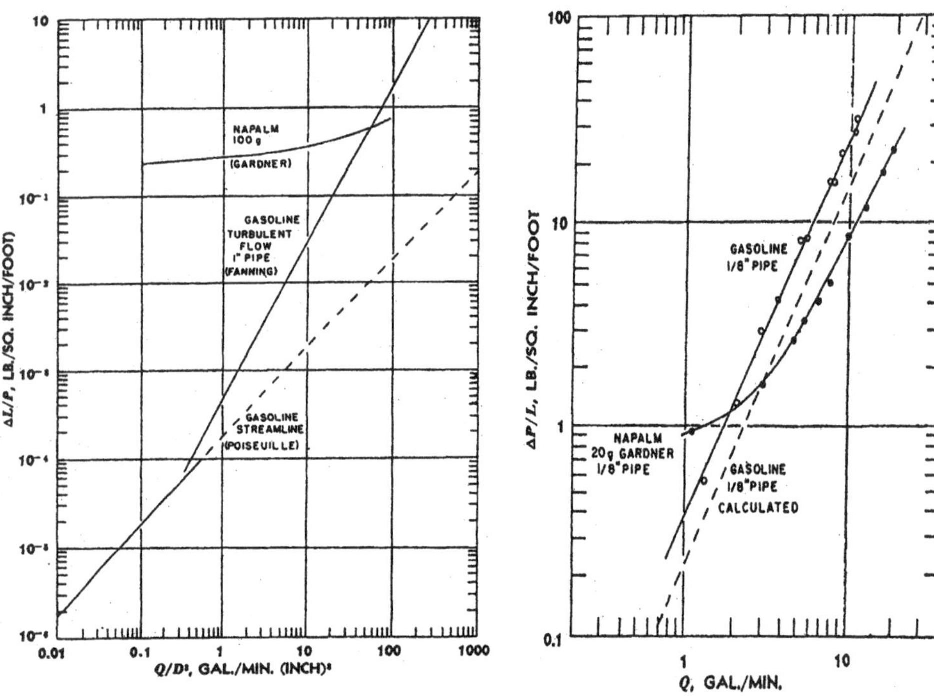

(a) Fig 2. Relation of pressure drop to flow. (b) Fig 3. Pressure drop of pure gasoline and dilute jelly.

Fig 7.9. Pipe flow data on gelled gasolines from Agoston et al (1954).

In 1971, Mysels gave a personal account of this work, describing *inter alia* the difficulties of the working conditions. This paper was part of a symposium on drag reduction at which both the current authors presented papers. The Proceedings were edited by P S Virk and J G Savins, and the latter pointed out that the literature on drag reduction had grown from one paper in 1953 to 400 in 1971.

Brian Toms gave his version of his discovery of drag reduction later (Toms 1977). At the time, he gave his address as the Department of Chemistry, University of Birmingham (where early work had been done by Garner's group in the 1939-45 war period). His remarks were 'personal and reminiscent'.

The original experiments of Toms were performed in 1946 and were presented at the 1st International Congress on Rheology in the Netherlands (Toms 1949). During the war years, he had worked on the mechanical degradation of polymers, and found that the actual breakage of molecules was important in solutions. This provided motivation for his work on turbulence, which he carried out in 1946 on solutions of polymethyl methacrylate (pmma) in monochlorbenzene "mainly because I happened to have enough of these materials for a lengthy investigation. (In those days, just after the war, one used whatever came to hand)".

Toms says of his experiments: "The really astounding thing was that in the critical region of concentration, and under turbulent conditions, a polymer solution clearly offered less resistance to flow, under constant pressure, than the solvent itself. This, of course, is the essence of drag reduction, and it was first noticed about the middle of 1946".

Toms states that Oldroyd (1949) suggested wall slip, rather than turbulence suppression as the cause, and when he (Toms) looked for the effect, an effective slip did seem to exist.

At the Rheology Congress in Scheveningen, he remembered that "only a small number of people expressed interest in my experimental results", and there were sceptics. By the mid-60s, people became interested and by the end of the decade there was a "small torrent" of activity, which agrees with Savins's paper count.

From the above sources, it is clear that the phenomenon of drag reduction was discovered independently at least twice, but Toms was the first to make his findings known to a scientific audience, in 1948.

The 'Toms effect' terminology is therefore based on a reasonable historical basis and, in the expanding research of the 60s, numerous researchers wrote of 'the Toms effect' (see, for example, Fabula 1965, Astarita 1965, Virk et al 1967). Specifically, Fabula, in an influential presentation at the 1963 International Congress on Rheology at Brown University, remarked: "There seems little doubt that the phenomenon is basically the same as that observed by Toms. Thus it is appropriate to recognize his precedence in this subject".

It is interesting that following the original studies of Toms and Mysels and co-workers in the 40s, the subject was not taken up again in earnest until the late 50s and early 60s. The initial interest was confined to drag reduction in *polymeric* liquids, particularly aqueous solutions of polymers like polyacrylamide, polyethelyne oxide and guar gum. Very soon, a consensus developed concerning the factors that enhanced drag reduction. Amongst these were the following (cf. Hoyt and Fabula 1964, Hoyt 1975, Lumley 1967, Hershey and Zakin 1967):

(i) Polymer linearity. Linear polymers were more effective than branched polymers.

(ii) High molecular weight, with 50,000 or higher required for effective drag reduction. A value of at least 10^5 was sometimes recommended.

(iii) Good solubility. 'Polymers dissolved in good solvents showed more drag reduction than in poor solvents'.

When these conditions were met, large reductions were often obtained, with polymer concentrations in the tens of parts per million range (see also Gadd 1966).

As the study of drag reduction blossomed, it was quickly realized that the phenomenon was not confined to *polymeric* systems, and Savins (1967, 1969) showed that micellar systems could also be effective drag reducers (see also White 1967a).

In the course of his work, Savins took out a patent on micellar-type drag reduction; he also discovered 'a stress-controlled' drag reduction phenomenon, which was peculiar to micellar systems (Savins 1967). In these systems, drag reduction was found to reach a maximum value at a critical wall shear stress, above which the drag reduction decreased. When the flow was slowed down, the drag reduction slowly recovered and reached a maximum at the same critical shear stress. This contrasted with the known behaviour in polymeric liquids, where the drag reduction tended towards a constant value, unless of course polymer degradation was in evidence, in which case the resulting lowering of drag reduction was irreversible.

To add to the rich diversity of the drag reduction phenomena, fibre suspensions were also found to exhibit the effect, although not to anything like the same degree as that found in polymeric liquids. So, for example, Bobkowicz and Gauvin (1965, 1967) observed drag reduction in aqueous suspensions of nylon fibres.

As the field expanded, there were many proposals for applications of the drag-reduction phenomenon. Ship drag was to be reduced, and some yacht races had to explicitly ban the use of soluble coatings containing drag-reducing agents. Fire-fighting vehicles were equipped with polymer dispensers, so as to take advantage of the suppression of droplets in a polymer-solution jet and the increased range of the jet. Stormwater drain capacity was, and still is, augmented by the addition of polymeric materials. Medical uses have even been contemplated, but the blood flow in many animals is, at best, only just into the turbulent regime. Submarines were to be quietened by the judicious addition of polymers, thus making them harder to locate with sonar. In short, the drag-reduction phenomenon was put to work, at least in a limited way, and it continues to be useful.

Not surprisingly, numerous attempts have been made to elucidate the mechanism(s) which cause drag reduction. We have already mentioned Oldroyd's (1949) wall-slip hypothesis, and although it may give a description of the drag-reduction phenomenon, it is not a satisfactory explanation, since it is clear that the *solvent* does not slip at the wall. Further, the viscosity of the dilute solutions used are not noticeably different from the solvent viscosity, so it is difficult to see how a slip mechanism can account for drag reduction.

The most obvious path has been to consider how the structure of turbulent flow is affected by the presence of the additives. Metzner and Park (1964) at first suggested that the ratio of normal stresses to shear stress would be sufficient to affect turbulence, but later Seyer and Metzner (1966, 1967) saw that the ratio of elongational stress to shear stress could be very large, and they suggested that this was a prime factor in suppressing turbulence. They specifically suggested that the rapidly-increasing elongational viscosity as embodied in the Upper-Convected-Maxwell bead-spring model (cf. Chapter 5) would be sufficient to damp down the turbulent fluctuations. This explanation, which clearly involves the viscoelastic properties of the solutions, has been elaborated and reinterpreted by others on numerous occasions since 1964. In 1969, Lumley reviewed earlier ideas; see also the review by Berman (1978).

Virk (1975) in an article entitled *"Drag reduction fundamentals"*, says that the "mechanism of drag reduction is still rather obscure..."; Berman (1978) said "a detailed explanation is still lacking". Attempts at more elaborate explanations continue (see, for example, Goldshtik et al 1982).

Finally, in this section, we refer to other phenomena which have been thought to have some relation to drag reduction. However, amongst these, Lumley (1969) makes the point that shear-thinning in *laminar* flow, which results in the pressure being less than that which would exist in a laminar flow at the zero shear viscosity level, does *not* count as drag reduction, nor does delayed transition to turbulence. At the same time, delayed transition is of interest in its own right and has provided the field with its own fascinating observations. Most notable has been the behaviour in high speed flow past solid bodies. For example, Merrill et al (1966) reported observations of drag *reduction* for flow past a flat plate, but sometimes drag *augmentation* for flow past a torpedo-shaped body. This is shown in Fig 7.10 which contains the percentage increase in terminal velocity when the bodies were placed on the bottom of a tank filled with an aqueous solution of polyethylene oxide and then accelerated to terminal velocity by a system of counterweights. Figure 7.10 clearly shows that, at the higher concentrations, the terminal velocity is decreased for the torpedo, but not for the flat plate, although at the lower concentrations there is drag reduction in both cases.

The resolution of the provocative experiments shown in Fig 7.10 can be assigned to a delayed transition to turbulence in the dilute polymer solutions, which, for the torpedo, delayed the transition to the narrower wake which is so characteristic of turbulent flow past bluff bodies (see also White 1966, 1967b). The answer to the different behaviours shown in Fig 7.10 is therefore related to Eiffel's paradox (see, for example, Birkhoff 1950).

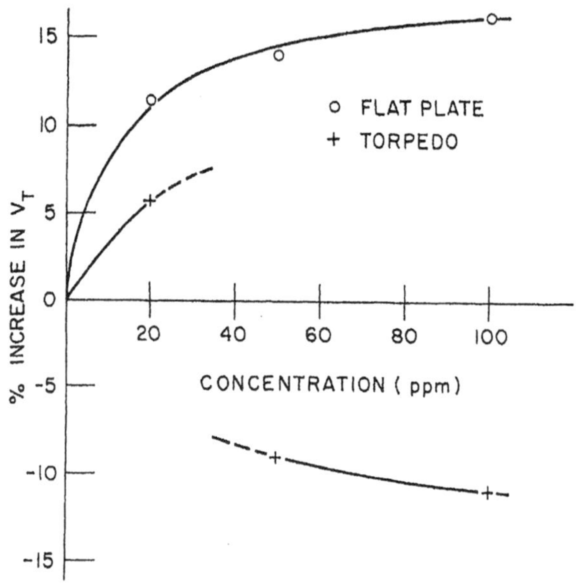

Fig 7.10. Experimental terminal velocity V_T data for aqueous solutions of polyox (WSR 301). Positive values of % increase in V_T imply drag reduction, negative values drag augmentation. (From Merrill et al 1966.)

7.8. D D Joseph

Dan Joseph was born in Chicago on March 26th 1929. He initially read Sociology at the University of Chicago, before moving to the Illinois Institute of Technology, where he majored in Mechanics and Mechanical Engineering. He obtained a PhD degree in 1963. In that year, Joseph moved to the University of Minnesota to begin a distinguished career, which has seen him move through the ranks to his present position(s) as Regent Professor and Russell J Penrose Professor of Aerospace Engineering and Mechanics.

Joseph's early research interests were centred in classical (Newtonian) fluid mechanics and Non-linear analysis and he quickly gained an international reputation in these fields. So much so that, when he entered the field of *non*-Newtonian Fluid Mechanics in the 70s, he brought with him a wealth of ability and experience. His research in this field is known for its careful experimentation, coupled with thorough theoretical analysis, the latter invariably staying within the security of the hierarchy equations of simple-fluid theory. This places a (severe) restriction on the experimental conditions of relevance, but at the same time permits confidence in the interpretation of the experimental results thus obtained.

Dan Joseph is a prolific publisher of research books and papers; he is an enthusiastic, entertaining, if sometimes lugubrious, lecturer, and he is in constant demand as a plenary speaker at international meetings. He has an impressive list of honours including membership of the National Academy of Sciences and the National Academy of Engineering. In 1993, he was awarded the Bingham Medal by the Society of Rheology.

Dan Joseph remains an energetic and popular elder statesman in the field of rheology; he clearly hates the prospect of old age. His jogging and marathon exploits are legendary.

7.9. Morton M Denn

Morton M Denn was born in Paterson, New Jersey on July 7th, 1939. He obtained a BSE, with highest honours, at Princeton University in 1961. As part of his undergraduate studies, he was introduced to rheology through a senior thesis, supervised by a young assistant professor named W R Schowalter. The relevant research involved the construction of a rather large 10 inch diameter cone-and-plate rheometer. From a limited set of data on a solution of polyisobutylene in decalin, it was concluded that normal stresses in a shear flow were real, something that was not generally accepted in 1960; at that time, some workers believed that normal stresses could be merely an 'end effect'.

In 1961, Denn moved to the University of Minnesota to do research on 'optimization in topologically-complex systems' and he was awarded a PhD degree in 1964. His Minnesota days were not entirely lost to rheology, since he made use of lecturing courses and seminars given or arranged by such distinguished scientists as A G Fredrickson, R Aris, J Serrin, L E Scriven and H Brenner (the latter being on sabbatical leave at Minnesota for a part of the time).

In 1964, W R Schowalter informed Denn that A B Metzner of the University of Delaware was looking for a postdoctoral worker, and Denn's appointment to the post began a distinguished association with the University of Delaware. Metzner's rheology laboratory was housed in an old Second World War quonset hut, and one of Denn's most vivid memory of the time was carrying out the 'hammer-whacking' experiment at Metzner's suggestion. Two small puddles were created on a concrete floor. One puddle was a very viscous Newtonian liquid, the other was a highly-elastic polymer solution known as 'super goop', a high molecular weight polyacrylamide in a mixture of glycerine and water. When the Newtonian liquid was hit, the fluid shot out as a spray; when the polymer solution was hit, it started to shoot out, but then snapped back. This led to the helpful Delaware approach to time scales and the significance of the Deborah number.

At Delaware, Denn quickly moved through the ranks and he became the Allan P Colburn Professor in 1977. He held this position until his move in 1981 to the University of California at Berkeley.

Denn has a forthright and articulate lecturing style and this is carried over to his interjections and contributions in open discussions at scientific meetings. By any yardstick, he has been a highly successful and influential chemical engineer and rheologist. He was Executive Editor of the A.I.Ch.E. Journal from 1985 to 1991, and since 1995, he has been Executive Editor of the Journal of Rheology, following in the footsteps of his one time Delaware colleague, A B Metzner. Denn was awarded the Bingham medal in 1986, the same year as his election to the National Academy of Engineering.

Morton Denn and his wife Vivienne are active in the local synagogue. They enjoy walking and are regulars at the opera and the theatre.

Chapter 8

Computational Rheology

8.1. Background and Overview

In this chapter, we trace the history of developments in the theoretical solution of rheological flow problems. By common consent, most of the problems of interest to rheologists are too complex to permit analytic solutions, and we shall quickly proceed to a detailed consideration of the relatively new field of *Computational Rheology.*

A number of authors have attempted to provide a convenient flow classification, so as to highlight the inevitable and inexorable move to *numerical* solutions of flow problems (see, for example, Crochet and Walters 1983a, Crochet et al 1984). These classifications usually include the following:

(i) Flows dominated by shear viscosity (see, for example, Bird et al 1987a, Ch4). In some situations of practical importance, the normal stress differences are low with respect to the shear stresses, and, even when fluids are reasonably elastic, the geometry of the flow may be such that fluid elasticity has only a minor influence on flow characteristics, lubrication-type situations being typical examples. In these circumstances, variable viscosity effects are usually important and need to be accommodated. This has invariably been accomplished by employing the so called Generalized Newtonian model (GNM) (equation 4.3 with $c_2 = 0$ and c_1 a function of the second invariant of $\boldsymbol{d}$). Existing techniques for solving Newtonian fluid mechanics problems have often been adapted with ease to meet the new challenge of a shear-dependent viscosity, the application of numerical techniques being especially helpful and efficacious in this regard (see, for example, Crochet et al 1984, Ch 9).

(ii) Slow flow (slightly elastic liquids). Here, the Weissenberg number (or the Deborah number) defined in §7.1 is small. In these circumstances, it has been possible to utilize the hierarchy equations of Coleman and Noll (cf. equation 4.8). The theoretical analyses have usually resolved themselves into perturbation problems, with perturbation either about a state of rest (in the case of slow flow) or about a known Newtonian solution (in the case of slightly elastic liquids) (see, for example, Walters 1979). The perturbation parameter has been the Weissenberg number or a suitable combination of the Weissenberg number and Reynolds number. Here, existing solutions to Newtonian flow problems have been extended without undue difficulty. The resulting equations are certainly more complicated in detail, but few new fundamental problems have emerged, since these have been masked by the perturbation nature of the solution method. Computational methods devised for the corresponding Newtonian problems readily lend themselves to the increased detail in the governing equations for the rheological problems.

There has been some reluctance, especially amongst applied mathematicians, to concede that the consistent use of the hierarchy equations requires not only that the flow is slow, but that it is also 'slowly varying'. This essentially rules out their use in most of the problems that are of industrial interest, flows near so called re-entrant corners[1] being obvious examples.

Note that the hierarchy equations of Coleman and Noll, such as the second order model, are *explicit*, in the sense that the stress is given explicitly in terms of kinematical tensors (see §4.4).

(iii) With the increasing capacity and speed of computers, attention in non-Newtonian fluid mechanics inevitably turned to complex flows involving highly elastic liquids (see, for example, Crochet and Walters 1983a,b), not only because of their practical relevance in such fields as polymer processing (cf. Crochet and Walters 1983b, Tanner 1983), but also because they have given rise to a number of difficult but stimulating problems of a fundamental nature, not encountered in classical fluid mechanics. By the mid-seventies, there was therefore a strong stimulus to apply numerical techniques to complex rheological flow problems and the field of computational rheology began to accelerate rapidly.

We may state the basic problems in computational rheology as being the numerical prediction of the behaviour of highly elastic liquids in complex flows, when the timescale of the flow process is short enough for the Weissenberg number (or Deborah number) to be relatively high. Such situations abound in industry, but it is only in recent years that the scientific know-how and, equally importantly, the size and power of available computing resources have been sufficient to meet the challenge. Accordingly, since the mid 1970's there has been a publication boom in the field; a boom that correlates directly with the increase in available computing power. It is therefore appropriate to include a discussion on general 'computer developments'.

8.2. Developments in Computer Power and Computational Techniques

Most areas of science have felt the impact of computing and rheology is no exception; indeed, it is a good example of the benefits of intensive computer application. We may recall that the first programmable electronic digital computer (ENIAC) ran in 1943, and that computers were uncommon in the early 1950s, the first commercially available machine being installed in 1954.

As an illustration of the situation in the 1950s, we recall an early centre of digital computation at the University of Manchester, UK, which had the second-generation Ferranti

[1]With hindsight, the term'salient corner', when referring to the solid boundary, would probably have been more appropriate.

Mercury computer. A practical run time of about 30 minutes maximum was imposed, not by bureaucratic fiat, but by the probability that an electronic valve (tube) would fail in that time.Papertape with punched holes (to signify a zero or a one) was used to instruct the machines, and it was possible to solve ordinary differential equations or compute definite integrals comfortably on this equipment (see, for example, Tanner 1962, 1963). The solution of field problems was also possible, but the shortage of memory (a few kilobytes) made life interesting.

By the mid 1960s, the switch to transistor technology had improved reliability and memory, and the solution of fluid mechanics problems by finite-difference methods (FDM) became commonplace. The finite-element method (FEM) was developed for computer use around this time (Turner et al 1956) and rapidly made for a revolution in stress-analysis techniques in solid mechanics, especially for the linear case. Application to fluid mechanics was a little slower; finite-difference methods were already well entrenched in this field. Furthermore, no variational principle was available for the Navier-Stokes equations, whereas all early finite-element methods depended on such a principle.

In duė course, potential-flow problems were solved by Martin (1968) and creeping-flow problems by Atkinson et al (1969); it was finally realized, following Szabo and Lee (1969), that a Galerkin approach to finite-elements was possible. A number of books began to appear (see, for example, Klein and Marshall 1968) with applications to rheology.

By 1970, computer speed and memory enabled large linear problems to be solved by FDM and FEM techniques and a good start was made on nonlinear field problems in rheology. One of the first papers to exploit this area was the computation of the flow of a power-law fluid in a square duct by D M Young and M F Wheeler (1964) of the University of Texas. Earlier, in 1961, Schechter had solved this problem using classical variational methods, where a set of functions was chosen to represent the solution in the domain of the problem. The choice of polynomials or Fourier series was common and a few (often three or fewer) terms were used in a truncated-series representation. Schechter (1961) represented the axial velocity component w by a series of the form:

$$w = \sum_{i,j} a_{ij} \sin \alpha_i x_1 \sin \beta_j x_2. \tag{8.1}$$

Later, this kind of global representation would be renamed 'spectral methods' by Gottlieb and Orzag (1977) and adapted by Beris et al (1987) and others (see, for example, Gervang et al 1992). To find the coefficients a_{ij} of the double Fourier series, Schechter used a variational principle, first derived by Pawlowski (1954) and afterwards rediscovered by several others, including Tomita (1959), Bird (1960) and Johnson (1960).

The Young and (Mary) Wheeler (1964) paper, which had connections with papers by J A Wheeler and Wissler (1965, 1966) and Middleman (1965), contained an alternative FDM numerical method.

The Wheeler and Wissler (1965) paper stated that "Schechter's method is impractical when highly accurate velocity profiles are needed". Schechter claimed about 1% accuracy, and Wheeler and Wissler aimed for an accuracy of 0.1%. While the assertion may be true if only a few terms are used in the variational/spectral approach, it is now conceded that the spectral method may give more accuracy for a given computational effort.

Wheeler and Wissler used 20×20, 40×40 and occasionally 80×80 meshes. For a shear thinning fluid, with the power law index less than 0.5, stability of the computation was a problem. This may have been due to the successive over-relaxation solution method used to solve the simultaneous equations.

We have dealt with this problem in some detail, since it brought the potential of the digital computer to the attention of several groups. While the simple FDM used was adequate, economical and appropriate at that time for reducing the partial differential equations to algebraic equations, it was not and is not the only method. We have already referred to the spectral method and the FEM, and other methods have also been used in computational rheology, including the boundary-element method (BEM) popularized by Cruse and Rizzo (1968). All these methods proceed from the (rheological) field problem and essentially reduce the field partial differential equations to a set of simultaneous, generally nonlinear, equations for nodal variables (or more general variables for some spectral methods). They are all (or nearly all) variants of a method of weighted residuals (see Finlayson 1972), whereby the field equations are written as

$$\boldsymbol{L}\boldsymbol{u} - \boldsymbol{f} = \boldsymbol{0}, \tag{8.2}$$

where $\boldsymbol{u}$ is a solution vector, $\boldsymbol{L}$ is an operator, and $\boldsymbol{f}$ is a driving function. $\boldsymbol{u}$ is then expressed as a sum of n modes and (8.2) is weighted on average over the space Ω of interest with respect to each mode g_n:

$$\int_{\Omega} (\boldsymbol{L}\boldsymbol{u} - \boldsymbol{f}) g_n dV = \boldsymbol{0}, \tag{8.3}$$

thereby producing a weak-form solution. The g_n depend on the method chosen; see the books by Finlayson (1972) and Crochet et al (1984) for further descriptions of the process.

Thus, by the early 1970s, methods of solution were available for tackling some nonlinear problems and computational facilities were improving. For example, the Control Data Corporation CDC 7600 had, in the early 1970's, a central memory of the order of 100k words and a speed of 10Mflops. Other computers were soon to become available of comparable speed.

The speed of computers has continued to increase, with at least a doubling every two years; for example, the jump from 10Mflops to 100Mflops took about 10 years, and, since 1990, the move to parallel computing has enabled an accelerated increase in speed and decrease in cost.

The general scheme for any numerical analysis follows Fig 8.1. The discretization in time is nearly always divorced from the rest of the problem, but the spatial discretization has brought forth great ingenuity. The inclusion of temperature variations, chemical reactions and other effects is now possible. These present no insuperable computational problems and will not be considered here.

8.3. The Distinctive Challenges of Computational Rheology

In many ways, developments in computational rheology have mirrored similar developments in conventional CFD and have been driven by them. At the same time, there are some distinctive challenges in the rheological setting, which certainly have no counterpart in the Newtonian case.

We have already implied that *explicit* constitutive models like the second order model (Eq. 4.9)are of very limited utility. In consequence, the constitutive equations employed have either been of the *implicit* differential type or have involved an equivalent integral formulation. For more than one reason, considerable prominence has been given to the Oldroyd B model given by equation (4.13) or the Upper Convected Maxwell (UCM) model (for which $\lambda_2 = 0$ in (4.13)). The main motivation was the relative simplicity of these models; a second consideration was the initial hope that such models would be adequate to describe the behaviour of the Boger fluids which were being used in some of the experimental programmes that developed alongside the numerical simulation studies.

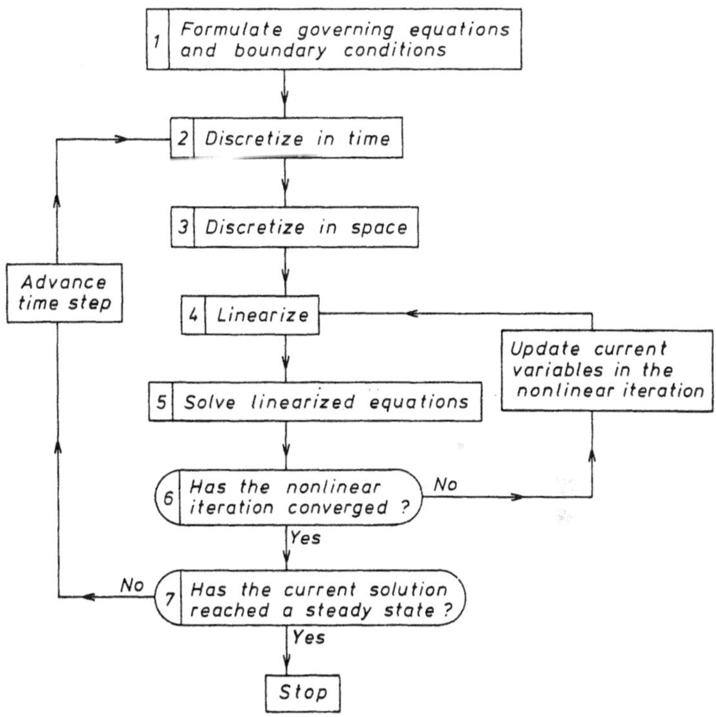

Fig 8.1. General scheme for numerical analysis. (From Crochet et al 1984.)

In retrospect, on both counts, the choice of model was *not* inspired. In the first place, the Oldroyd-B and UCM models introduce serious numerical challenges that are somewhat alleviated by the use of more complicated constitutive models. These models, unlike the Oldroyd-B and UCM models, usually have a shear-thinning viscosity behaviour; they also cap the extensional viscosity, so that it does not become infinite at a finite stretch rate. Both these features have beneficial consequences in the numerical simulation.

In the second place, it was slowly realized that associating such simple models with Boger fluids was too optimistic, if not naive (see, for example, Rallison and Hinch 1988). So, paradoxically, the move to more realistic and more complicated constitutive models for Boger fluids and other polymeric liquids had beneficial spin-offs.

In order to highlight the distinctive problems encountered in computational rheology, it is useful to display the governing equations for a steady two dimensional flow for the UCM model.

Let u and v be the velocity components in the Cartesian x and y directions, respectively. The (non-dimensional) governing equations for the UCM model can be written in the form (cf. Crochet and Walters 1983a):

$$T_{xx}\left[1-2W_e\frac{\partial u}{\partial x}\right]+W_e\left[u\frac{\partial T_{xx}}{\partial x}+v\frac{\partial T_{xx}}{\partial y}\right]-2W_eT_{xy}\frac{\partial u}{\partial y}=2\frac{\partial u}{\partial x}, \tag{8.4}$$

$$T_{yy}\left[1-2W_e\frac{\partial v}{\partial y}\right]+W_e\left[u\frac{\partial T_{yy}}{\partial x}+v\frac{\partial T_{yy}}{\partial y}\right]-2W_eT_{xy}\frac{\partial v}{\partial x}=2\frac{\partial v}{\partial y}, \tag{8.5}$$

$$-W_eT_{xx}\frac{\partial v}{\partial x}-W_eT_{yy}\frac{\partial u}{\partial y}+W_e\left[u\frac{\partial T_{xy}}{\partial x}+v\frac{\partial T_{xy}}{\partial y}\right]+T_{xy}=\frac{\partial u}{\partial y}+\frac{\partial v}{\partial x}, \tag{8.6}$$

from the constitutive equations;

$$f_x-\frac{\partial p}{\partial x}+\frac{\partial T_{xx}}{\partial x}+\frac{\partial T_{xy}}{\partial y}=R_e\left[u\frac{\partial u}{\partial x}+v\frac{\partial u}{\partial y}\right], \tag{8.7}$$

$$f_y-\frac{\partial p}{\partial y}+\frac{\partial T_{xy}}{\partial x}+\frac{\partial T_{yy}}{\partial y}=R_e\left[u\frac{\partial v}{\partial x}+v\frac{\partial v}{\partial y}\right], \tag{8.8}$$

from the stress equations of motion; and

$$\frac{\partial u}{\partial x}+\frac{\partial v}{\partial y}=0 \tag{8.9}$$

from the conservation of mass equation. In these equations, T_{ik} is the (non-dimensional) extra-stress tensor, $(f_x,\ f_y)$ are the components of the body force per unit mass, p is the pressure and R_e and W_e are the Reynolds number and Weissenberg number, respectively; the latter involves the relaxation time λ of the UCM model. For the *integral* equivalent of the UCM model, the constitutive equations are given by

$$T_{ik}(\,t)=\left(\eta/\lambda^2\right)\int_{-\infty}^{t}\exp\left[-(t-t')/\lambda\right]\left[C_{ik}^{-1}(t')-\delta_{ik}\right]dt', \tag{8.10}$$

where C_{ik}^{-1} is the Finger tensor defined by

$$C_{ik}^{-1}=\frac{\partial x_i}{\partial x'_m}\frac{\partial x_k}{\partial x'_m}, \tag{8.11}$$

and x'_i is the position at time t' of the fluid element that is instantaneously at the point x_i at time t. In a two-dimensional problem, one of the ways of obtaining the so called displacement function x'_i is to solve the equations (Oldroyd 1950):

$$\frac{\partial x'}{\partial t}+u\frac{\partial x'}{\partial x}+v\frac{\partial x'}{\partial y}=0, \tag{8.12}$$

$$\frac{\partial y'}{\partial t} + u\frac{\partial y'}{\partial x} + v\frac{\partial y'}{\partial y} = 0. \tag{8.13}$$

It is immediately apparent that, in the case of the *differential* UCM model, there are six equations in the six unknowns u, v, p, T_{xx}, T_{xy}, T_{yy}.

For the integral model, there are essentially five equations for the five unknowns u, v, p, x', y'.

Clearly, whether one uses the differential or integral forms of the simple UCM model, there are significant complications to the Newtonian case, where the Navier-Stokes equations would result in *three* equations in the three unknowns u, v, p.

The first successful attempts to solve rheological flow problems for an implicit differential equation were those of Perera and Walters (1977a,b) using the FDM and by Kawahara and Takeuchi (1977) who employed the FEM, using a very coarse mesh. Other early work by Finlayson and his students used a collocation process (Chang et al 1979).

Solutions to the integral equivalent of the UCM had to wait until the 1980s. Viriyayuthakorn and Caswell (1980) and Bernstein and Malkus (see, for example, Bernstein et al 1981, Bernstein and Malkus 1982 and Malkus and Bernstein 1984) employed the FEM and Court et al (1981) successfully used the FDM.

These papers, written in the late 70s and early 80s, set the stage for the dramatic increase in activity in the following years, which coincided with the emergence of a number of dominant international research groups, most notably the Louvain-la-Neuve, Belgium, group led by **M J Crochet**, the M.I.T. team under R C Armstrong[2] and R A Brown[3] and the Sydney group of R I Tanner and N Phan-Thien.

There were numerous other problems that could not have been anticipated at the birth of computational rheology, most notably the so called High Weissenberg Number Problem, which we shall describe in the next section.

8.4. Progress is Made

Early work on the finite-difference applications is described in the book by Crochet et al (1984). Often, the pressure was eliminated by using (in two-dimensional flows) a stream-function/vorticity formulation. Whilst this enabled progress to be made in solving the problem for implicit differential models, it proved difficult to recover the pressure field at the end of the computations. This was not unique to rheology of course, since the problem also existed in similar Newtonian flow simulations.

---

[2]Robert C Armstrong was born in Baton Rouge, Louisiana in May 1948. As an undergraduate, he studied at the Georgia Institute of Technology, where he graduated in 1970 with the highest honours in Chemical Engineering. He then did his graduate work at the University of Wisconsin, Madison, working under the supervision of Professor R B Bird; he was awarded the PhD degree in 1973. Armstrong immediately joined the Faculty of the Massachusetts Institute of Technology and he has remained there ever since. Since 1996, he has been Head of the Department of Chemical Engineering.

[3]Robert A Brown was born on 22nd July 1951 in San Antonio, Texas. After obtaining his Bachelor of Science and Master of Science Degrees at the University of Texas, he obtained his PhD (1979) from the University of Minnesota in the field of chemical engineering. He then went to the Department of Chemical Engineering at the Massachusetts Institute of Technology, rising to the rank of Professor in 1984. In 1989 he became more involved in administration, being first department head and then moving to his current position as Dean of Engineering. He is a member of the (US) National Academy of Engineering.

In order to recover the elliptic operator for the vorticity, which was present in the Newtonian case, Perera and Walters (1977a) introduced a useful transformation, which involved the splitting of the extra-stress tensor T_{ik} in the following way:

$$T_{ik} = S_{ik} + 2\eta d_{ik}, \tag{8.14}$$

where η is a reference viscosity. The problem was then recast with S_{ik} as the unknown variable. This splitting has been used by others in later studies, sometimes without reference to the Perera and Walters work.

The second-order fluid permitted similar transformations, following the discovery of several useful theorems (see, for example, Tanner and Pipkin 1969), but the second-order model was (and is) not a simple model with which to compute, nor is it a realistic model for general use as we have already indicated, and its use has virtually disappeared as the field has progressed.

While these early FDM studies were of great interest and were suited to small computers, their convergence was limited (with a typical upper bound on W_e in the region of 0.5) (cf. Crochet et al 1984). Some unsteady flow problems were also attempted using the FDM, and, as early as 1973, Townsend successfully solved an unsteady problem involving just one space variable. Some time was to elapse before an unsteady problem was solved for the case of two space variables (see, for example, Townsend 1984).

Before we leave the FDM, we refer to the early work of Crochet and Pilate (1976) who employed the explicit second-order model. Soon, the Louvain la Neuve work was to embrace the FEM and employ implicit differential models.

At first it was considered that a variational principle was needed to apply the FEM, but, in 1969, Szabo and Lee showed that a Galerkin approach was feasible, so this restriction was removed. Tanner (1973) and Nickell et al (1974) later produced the first free-surface analyses of extrusion and demonstrated the advantages of the FEM in terms of geometrical flexibility and the application of mixed fixed-and free-boundary conditions, which so often occur, for example, in polymer processing applications. The following year saw (Tanner et al 1975) the first nonlinear rheological analysis of this type, using an inelastic GNM of the power-law type. Subsequently, many more workers began to enter the field.

It was soon discovered that the analysis of the second-order model by the FEM was no easier than by any other method. A short paper on the extrusion of a plane sheet by Reddy and Tanner (1978) showed that, with a fixed, relatively coarse, mesh, a Weissenberg number of about 1 could be reached. It was not possible to solve the corresponding axisymmetric flow problem at any significant W_e value with the computational method employed.

So, by the late 70s, many workers were becoming frustrated by what quickly became known as the 'High Weissenberg Number problem' (HWNP) - there was an upper limit on W_e, written $W_{e_{crit}}$, above which the numerical algorithms failed. In 1984, Crochet et al referred to 'the outstanding problem in the numerical simulation of non-Newtonian flow' and made the following observations:

(i) A limit on W_e is found in all published work.

(ii) Minor changes in the constitutive equation and/or the algorithms employed could lead to higher limiting values of W_e. However, such improvements do no more than delay the breakdown process.

(iii) As W_e approaches $W_{e_{crit}}$, it is often (but not always) observed that spurious oscillations appear in the field variables.

(iv) Mesh refinement and 're-entrant corner' strategies affect $W_{e_{crit}}$, but it is 'difficult to discern an overall consistent trend in published work'.

The HWNP was discussed informally by a small group of attendees at an IUTAM meeting held in Louvain la Neuve in 1978. This turned out to be an immensely influential gathering, since it set in motion a continuing series of International Numerical Simulation Workshops. These have been of inestimable value in giving the field a focus and a forum.

The first Workshop, organized by Bruce Caswell, was held in Providence, Rhode Island, in 1979. This was followed by a more formal Workshop at Ross Priory, Scotland (see Journal of Non-Newtonian Fluid Mechanics (JNNFM) 1982, **10**, 1-118), with a return to New England in June, 1983, for a meeting at Lake Morey, Vermont (see JNNFM 1984, **16**, 1-209).

Attendance reached a peak at the Workshop held at Spa, Belgium, in June 1985 (see JNNFM 1986, **20**, 1-339). Over 70 were in attendance at this Workshop, most of whom are included in the group photograph shown in Fig 8.2.

A fifth Workshop was held at Lake Arrowhead in June 1987 (see JNNFM 1988, **29**, 1-443) and a sixth took place at Hindsgavl, Denmark, in 1989. The report of the Lake Arrowhead meeting stated "... there was a general sense that convergent calculations can now be obtained in a range that overlaps meaningful experiments"; also, it was here that benchmark problems were proposed.

The Hutchinson Island, Florida, Workshop in February 1992 (see D G Baird and M Renardy, JNNFM 1992, **43**, 383-385) was poorly attended, with only 35 participants and interest in the subject had temporarily peaked. However, attendance began to improve at the October 1993 workshop on Cape Cod (see R A Brown and G H McKinley, JNNFM 1994, **52**, 407-413), and, in the 9th Workshop held at Rossett, North Wales in April 1995 (see B Caswell, JNNFM 1996, **62**, 99-110), attendance again reached 75. The 10th Workshop was held at Ocean City, Maryland, in May 1997, and there is every indication that research activity in the field is buoyant, with a significant input from young rheologists.

Solutions to the HWNP have gradually evolved since 1977. An early idea of importance was that of Crochet and Keunings (1980), who used a system of coupled equations, drawing on experience gained previously from Newtonian problems. Here, the unknown pressure, velocity and stress components were solved simultaneously, using Gaussian elimination. This led to a considerable advance in the critical Weissenberg number attainable (Crochet and Keunings 1982). The solution method was coupled to a Newton-Raphson scheme, which enabled accurate solutions to be made, once divergence had been avoided. The relevant schemes (known as MIX-schemes) were discussed at the 3rd Workshop at Lake Morey in 1983. At this time, a clear recognition of the source of the convergence problems began to appear. The *hyperbolic* nature of constitutive equations of the differential type was pinpointed at this meeting, and it was recognized as an important factor

Fig 8.2. Participants at the 4th International Workshop on Numerical Methods in Viscoelastic Flow; Spa, Belgium; June 3-5, 1985. (From J non-Newtonian Fluid Mechanics **20**, 1986, 23.)

1. C Guillope
2. S Kim
3. D V Boger
4. K Walters
5. J Son
6. M Renardy
7. M M Denn
8. R I Tanner
9. B Caswell
10. M Sabbah
11. M J Crochet
12. G D Richard
13. C L Tucker
14. J R A Pearson
15. S M Muller
16. M Apelian
17. J J Lumley
18. R A Brown
19. F Dupret
20. A R Davies
21. H Thieu
22. S M Dinh
23. R T Mifflin
24. O Hassager
25. R Keunings
26 T E R Jones
27. J M Marchal
28. W P Raidford
29. S Dupont
30. M Kim-E
31. F Dijksman
32. R G Larson
33. J C Saut
34. H vanWijngaarden
25. R C Armstrong
36. B Debbaut
37. J J van Schaftingen
38. T N Phillips
39. P W James
40. R S Jones
41. S Karrila
42. A S Lodge
43. Ph Le Tallec
44. W R Schowalter
45. J M Piau
46. M F Webster
47. R B Bird
48. E Mitsoulis
49. D S Malkus
50. G Leal
51. J V Lawler
52. G Marrucci
53. V Delvaux
54. J Baranger
55. H Holstein
56. J Dheur
57. M Hulsen
58. H Meijer
59. B Finlayson

in the stability problem. The difficulties with the stability of numerical schemes for the explicit second-order model was also pointed out around this time (Tanner 1982).

The Spa Workshop in 1985 witnessed a clear resolution to concentrate on a limited number of standard problems, in order to measure progress, since at that time no clear resolution of the HWNP had yet emerged. Early work had discussed die swell (Fig 8.3a) and the 4:1 contraction (Fig 8.3b) problems, but both contain points that lead to singular stress behaviour, even in the Newtonian case, and it was known that this led to a destabilization of the numerical schemes. Accordingly, the sphere-in tube problem (Fig 8.3c) was introduced at the Spa meeting. In this (nondimensional) problem, a sphere, radius unity, moves through a long tube of radius two; the UCM constitutive model was the preferred choice. The advantage of this problem was that it had no sharp corners and no free surfaces. Other problems suggested were the 'wiggly tube' (Fig 8.3d) and the eccentric rotating 'bearing' (Fig 8.3e). Subsequent workshops contained numerous attempts at solving these bench-mark problems as the extensive Workshop reports and proceedings testify. The sphere-in-tube problem has proved to be unexpectedly difficult and we shall give a brief history of progress on this problem later in this chapter.

In 1985, Yoo and Joseph discussed the 'change of type' in the governing equations, from elliptic/hyperbolic to completely hyperbolic for the kind of problems we are discussing. Later, Joseph et al (1987) applied the ideas to explain delayed die swell (cf. §7.3). For the Oldroyd B model (with $\lambda_2 \neq 0$) in the limit of creeping flows (i.e. $R_e \to 0$), change of type does not occur (see, for example, Crochet 1989).

Beris et al (1984, 1986) introduced the eccentric journal-bearing problem (Fig 8.3e) and here the Deborah number D_e is $\lambda\Omega$, where λ is the characteristic relaxation time and Ω the rotational speed of the inner cylinder. A D_e of 3 was reached with difficulty. Debbaut and Crochet (1986) used the Phan-Thien Tanner model in their simulations and stated that: "the loss of convergence is of purely numerical origin". This statement (made in connection with flow into a 4:1 contraction) is of importance, since much speculation about non-existence of solutions (especially *steady* solutions) had previously been aired.

In 1986, Luo and Tanner used a streamline-element scheme in conjunction with a realistic integral model for a polyethylene melt, which had been studied extensively by a IUPAC working party. A reasonable agreement with experimental values of die swell was shown at the low to medium rates of extrusion (Luo and Tanner 1988). At the time, it seemed likely that slip at the exit (see Phan-Thien 1987) was the cause of the lack of agreement at higher extrusion rates. Some time later, Goublomme et al (1992), using a similar computational technique, found a similar problem in modelling the behaviour of high density polyethylene at high extrusion rates. In this case, the excessive elasticity embodied in the integral constitutive equation was shown to be a likely cause of the disagreement (see also Crochet and Walters 1993).

So far as integral models are concerned, the iterative (Picard) methods initiated by Luo (see Luo and Tanner 1986, 1988, Luo and Mitsoulis 1989) remain the main methods of computation. We note that the HWNP has been alleviated here by the use of 'softer' constitutive equations.

By contrast, the differential constitutive models have generated many ingenious solutions to the HWNP. For example, Beris et al (1987) had much success with spectral methods in the eccentric-bearing problem. For small eccentricity, a Deborah number of

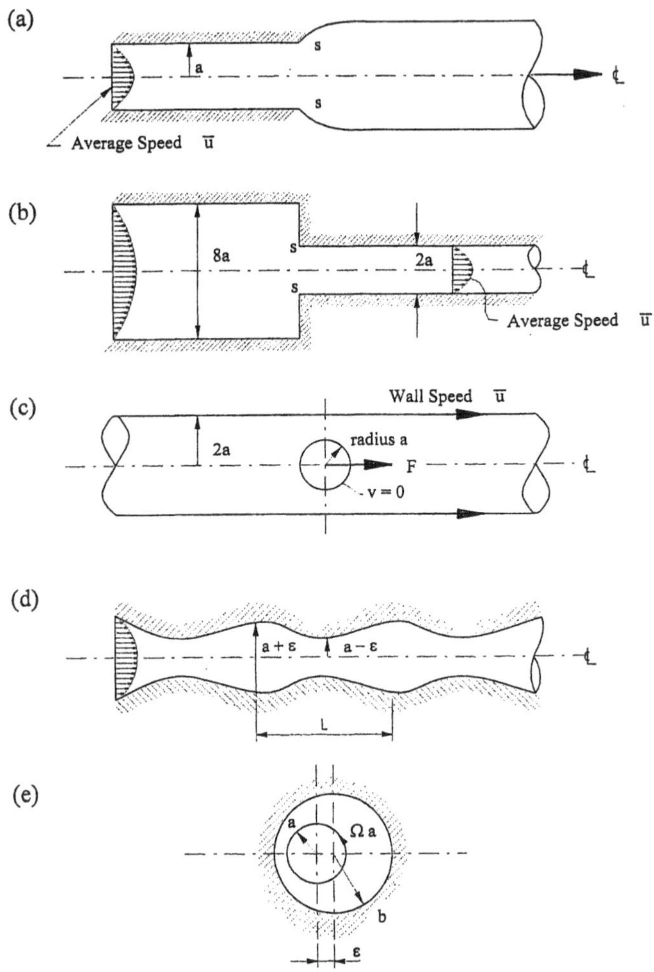

Fig 8.3. Five Standard Test Problems.

(a) Extrusion (plane or axisymmetric). Weissenberg Number $W_e = \lambda\bar{u}/a$, where λ is a relaxation time. Note points of singular stresses S, which are difficult to capture numerically.

(b) 4:1 Contraction flow (plane or axisymmetric): the points S cause problems.

(c) 2:1 Tube/sphere problem; $W_e = \lambda\bar{u}/a$. No singular points. Drag force F leads to definition of drag coefficient $K = F/6\pi\eta_0\bar{u}a$, where η_o is the zero-shear rate viscosity.

(d) "Wiggly" tube problem; diameter varies sinusoidally with distance along tube. Amplitude (ϵ) and wavelength (L) are parameters in problem.

(e) Eccentric rotating cylinder ("bearing") problem. Two dimensionless criteria are $W_e = \lambda\Omega a/(b-a)$ and $De = \lambda\Omega$; the eccentricity ratio $\epsilon/(b-a)$ is also important. (Ω is the angular speed of the inner cylinder).

90 was reached; this set a standard for convergence and accuracy. However, Spectral methods were (and are) awkward to implement in many complex geometries.

The Louvain-la-Neuve group had a brief flirtation with a Streamline Upwinding (SU) method (see, for example, Marchal and Crochet 1987). However, some of its deficiencies were exposed (see, for example, Tanner and Jin 1991, Legat and Marchal 1993) and it fell out of favour, especially since other methods were found to be more economical.

In 1988, King et al published the EEME method that was very clever in altering the variables to produce an "Explicitly elliptic momentum equation". The method was extended by Jin et al (1991) to a Phan-Thien Tanner model.

Many other approaches were tried. Tran-Cong and Phan-Thien (1988) used the boundary-element method at modest Weissenberg numbers to produce the first design of a die for a specific extrudate shape. Fortin and co-workers (1989, 1992) pioneered the use of iterative-solver schemes and decoupled methods using the Lesaint-Raviart scheme, which introduces damping and stability.

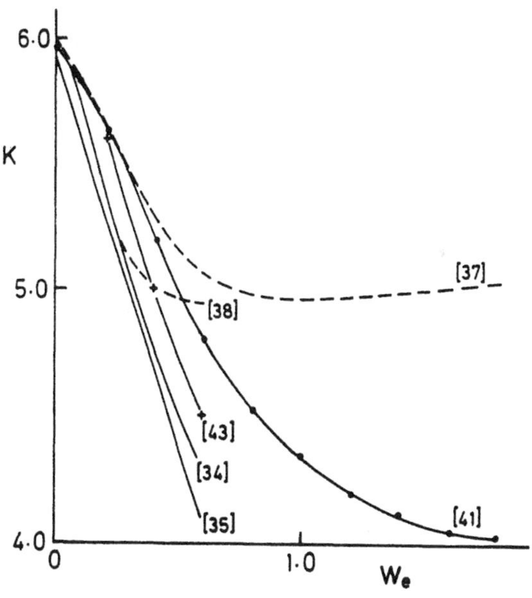

Fig 8.4. A plot of the drag correction factor K as a function of Weissenberg number W_e for a UCM model, for the case when the radius of the cylindrical container is twice that of the sphere. The numbers correspond to references of the published work on the problem that existed at the time. (From Walters and Tanner 1992.)

Later, the MIT group led by Brown and Armstrong produced the EVSS (Elastic Viscous Stress Splitting) scheme (Rajagopalan et al 1990), which seems to possess acceptable characteristics. EVSS is a fully coupled scheme and therefore requires considerable computer time. Sun et al (1996) have recently introduced a very stable uncoupled variant (AVSS/SI), which is fast. Khomami et al (1994) introduced the so called h-p methods. Guénette and Fortin (1995) and Baaijens (1994) have also produced EVSS-type schemes with good stability. Carew et al (1993) used a Taylor-Galerkin technique; Luo (1996) has introduced a pseudo-time algorithm. It is therefore evident that an enormous amount of ingenuity has been expended in developing schemes. We will now compare the utility of some of them by discussing the 1:2 sphere: tube test problem using the UCM model (Fig 8.3c).

Figure 8.4 from Walters and Tanner (1992) shows the considerable disarray that existed in 1990. No trustworthy results were available beyond a W_e of about 1.0 and many of the methods were showing problems long before that. By 1993, the picture looked much better (Fig 8.5) (see, for example, Crochet and Walters 1993, Rasmussen and Hassager 1993).

Using EEME and EVSS methods, Lunsmann et al (1993) produced consistent results up to a W_e of 1.5. Jin et al (1991) also obtained similar results from an EEME method. Later, Fan and Crochet (1995) used an EVSS method and were able to avoid divergence up to a W_e of about 2.1. Baaijens et al (1996) were able to reach a W_e of 2.5; Sun et al (1996) reached 2.6 and Luo (1996) published results up to 2.8. The simulations for $W_e > 1.6$ are again showing some divergence, but the picture is constantly changing and, with proper mesh refinement, the various curves are showing more agreement. Indeed, the 1991-1993 experience is being enacted again at the higher W_e values!

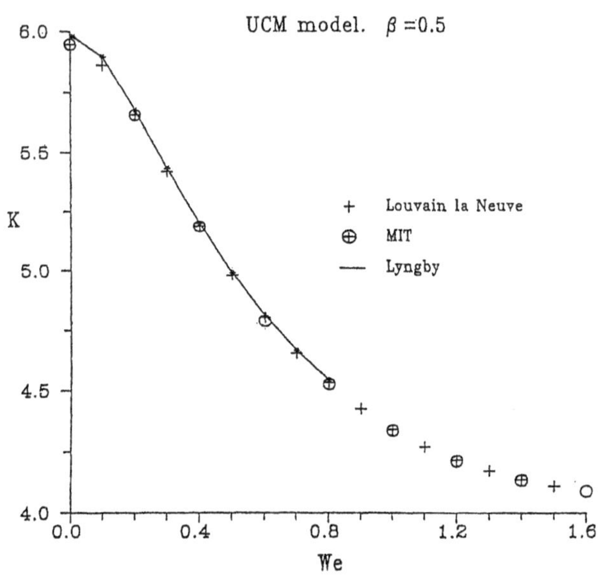

Fig 8.5. A plot of the drag correction factor K as a function of the Weissenberg number W_e for a UCM model, for the case when the radius of the cylindrical container is twice that of the sphere. Simulations obtained by three research groups have been taken from Crochet and Walters (1993). Numerous other groups have since published data which are in agreement with the results shown.

A number of important points can be made at this juncture. First, a serious suggestion that there was some kind of stability limit at $W_e = 1.6$ is now seen to be unfounded. Secondly, the HWNP, at least for a steady two-dimensional 'smooth' problem is now under control. To reach yet higher W_e values, it will be necessary to resolve the thin stress boundary layers that are an important feature of the simulations. This will require further mesh refinement; but the problem is now at a stage where there are no in-principle blocks in the way. Renardy (see, for example, Renardy 1997a, 1997b) has shown that the thin stress boundary layers are of order W_e^{-1} in the UCM fluid, and of order $W_e^{-1/3}$ for the PTT model.

In the current discussion, *three*-dimensional simulations have been conspicuous by their absence. Most of the available results have been very expensive to run and so a number of alternative paths for three-dimensional problems have emerged. In this connection, we may mention the finite-volume method, which has been employed in two dimensions by Song and Yoo (1987), Hu and Joseph (1990), Yoo and Na (1991), Darwish et al (1992) Sasmal (1995) and Huang et al (1996). The method has been employed in a three-dimensional problem by Xue et al (1995). These methods seem very fast and can be used on work stations. The future will be watched with interest. One thing is clear: The 'computational rheology' aspect of our 'history' will be quickly in need of updating.

8.5. Direct Simulation of Polymer Flow

The previous section has dealt with the use of computers to solve mathematical problems for polymeric liquids using the continuum approach. Another possibility is to use computing for more direct simulation taking 'microstructure' into account explicitly. Whilst this idea was tried in the 1970s, the lack of speed and memory frustrated early attempts. However, in 1978, with the growth of computer power, simulations were employed by Fixman (1978), Ermak and McCammon (1978), Pear and Weiner (1979) and Helfand et al (1980) to study the relaxation and flow of polymer chains. Since then, there has been an explosion of activity, not only for polymeric liquids, but also in colloidal suspensions and in flows with reactions.

Simulations for the Green-Tobolsky-Lodge-Yamamoto network model were made by Petruccione and Biller (1988a,b) and Biller and Petruccione (1990). A volume edited by Elizabeth Colbourn (1994) describes some of the wide field now being attacked in this way: computer-aided molecular design; molecular dynamics; modelling of amorphous polymers; Monte Carlo studies of collective phenomena in dense polymer systems; crystalline polymers; failure mechanisms networks and bio-polymers. Öttinger (see, for example, Laso and Öttinger 1993, Öttinger 1996) has championed the calculation of viscoelastic flow using molecular models via the so called CONNFFESSIT approach. Kremer and Grest (1990) have discussed the interaction of polymers with walls. In the mid 1980s, Brownian Dynamics was developed for suspensions. The particles are damped (Langevin equations here in place of Newton's for molecular dynamics (MD)) and many papers on dispersions now use these or similar methods (e.g. Heyes and Melrose 1993, Grassia et al 1995). Keunings (1997) has carried out a study of Brownian dynamics in non-Hookean dumbbells and has shown the strengths and weaknesses of earlier approaches based on simplified linearizations of the Peterlin type (cf. Chapter 5).

Heyes (1992) gives a short history of molecular simulation applied to rheology, emphasizing that the declining cost of computation makes it even more possible to use direct simulation methods. By direct simulation, we imagine a group of model particles, which could be small molecules or even macromolecules, that are set up in the computer, usually in a box with porous walls, so that periodic arrays of molecules can be dealt with in repeated images. The molecules are allowed to move around under Newton's laws, and stresses and rates of deformation can be computed for the models. This procedure leads to Molecular Dynamics (MD) simulations. Brownian Dynamics (BD) deals with the Langevin equations (Chapter 5) and is more suited to dealing with molecules in solution.

According to Heyes (1992), the first appearance of molecular simulation in rheology occurred in the 1960's, when the Green-Kubo formula was introduced to compute the zero-shear rate viscosity, without needing to shear the sample! In 1975, Ashurst and Hoover actually did shear the sample between walls, but later Evans and Morriss (1988, 1990) used the Green-Kubo method to deal with non-Newtonian flow, including normal-stress effects. Shear thinning in n-butane has recently been investigated; the main problems at the moment are the small molecules (hexadecane, for example) and the very large shear rates ($\dot{\gamma} \sim 10^{12}s^{-1}$) at which one sees shear thinning. At such rates, "thermostatting" is necessary and discussion on best methods is still progressing.

Evans (1988) has computed normal-stress effects in disk-like fluids in shear flow; the effects of walls have been considered by Jabbarzadeh et al (1997); as long ago as 1986, Heyes reported shear thinning in a Lennard-Jones potential liquid.

While it is clear that, at present, these methods are not generally competitive with continuum-based methods, they may become so very soon, at least for research purposes. The growth of parallel computing is clearly important here and will assist in the 'computer-physics' phenomenon.

8.6. M J Crochet

Marcel Crochet was born in Brussels, Belgium, in November 1938. His early academic training was in Electrical and Mechanical Engineering at the University of Louvain, where he graduated in 1961. He then carried out postgraduate work in Applied Mechanics at the University of California, Berkeley, where he obtained an MSc in 1964 and a PhD in 1966; his research supervisor was the well known applied mathematician, P M Naghdi.

Crochet moved back to Belgium in 1966 with an appointment as Assistant Professor at the Université Catholique de Louvain, Louvain-la-Neuve. He was appointed full Professor in 1975.

Crochet's early research was in *solid* mechanics, reflecting the influence of Paul Naghdi. However, a gradual transition of interests took place in the early 70's, and most of the subsequent research has involved viscoelastic *fluids*, beginning with some innovative research into nonisothermal problems, before the concentration on numerical techniques for viscoelastic flow calculation. This research was to play an important international role in the rapidly developing field of computational rheology. Crochet's early work in this field involved the finite-difference method, but he quickly embraced the finite-element technique, which had been originally developed within a solid-mechanics framework. He adapted the method with consummate skill, aided by a ready supply of first class graduate students, and he quickly became the leading authority in the field.

1982 saw the creation and first industrial implementation of the well known POLYFLOW program for the numerical simulation of Newtonian and non-Newtonian flows and, in 1988, the Polyflow s.a. Corporation was founded, with Crochet as the chief executive officer, a post he held until 1995.

Marcel Crochet is a superb lecturer, who has been in constant demand as a plenary speaker at international conferences. His organizational skills are impressive and many recall the truly excellent IUTAM meeting he organized at Louvain-la-Neuve in 1978 (see Journal of non-Newtonian Fluid Mechanics Vol 5). He was co-chairman of the Eleventh International Congress on Rheology held at Brussels in 1992.

Crochet has a penchant for good food and wine and is an excellent host, a role which he often fills in tandem with his charming wife, Brigitte. They have one son and one daughter.

In 1994, Marcel Crochet was awarded a Gold Medal by the British Society of Rheology and in 1995 he was elected Rector at the Université Catholique de Louvain, the top administrative post in that prestigious institution.

Appendix 1
Rheometrical Functions (Notation)

As the science of Rheology evolved, a plethora of different notations emerged; this was nowhere more in evidence than in the case of the various rheometrical functions. In an attempt to rationalize the situation, the American Society of Rheology provided a recommended notation (see, for example, Dealy 1984). This exercise has been largely successful.

So, for example, in the case of a steady simple shear flow with velocity components in a Cartesian coordinate system (x, y, z) given by (cf. Fig 1.1)

$$v_x = \dot{\gamma}y, \; v_y = v_z = 0,$$

where $\dot{\gamma}$ is a constant shear rate, the corresponding components of the symmetric stress tensor σ_{ik} are written as

$$\left.\begin{aligned} \sigma_{xx} - \sigma_{yy} &= N_1(\dot{\gamma}), \\ \sigma_{yy} - \sigma_{zz} &= N_2(\dot{\gamma}), \\ \sigma_{xy} = \sigma &= \dot{\gamma}\eta(\dot{\gamma}), \end{aligned}\right\}$$

with the other σ_{ik} zero. σ is known as the shear stress, η as the shear viscosity and N_1 and N_2 are the first and second normal stress differences, respectively. η_1, N_1 and N_2 are functions of $|\dot{\gamma}|$, and so are even functions; σ is an odd function of $\dot{\gamma}$.

Equivalent flow fields in other coordinate systems are known as viscometric flows (the terminology being due to Coleman 1962a, see also Coleman et al 1966) and the associated material functions σ, N_1 and N_2 are now generally referred to as the 'viscometric functions'.

In the case of a small amplitude oscillatory motion with Cartesian velocity components given by

$$v_x = \epsilon\omega y e^{iwt}, \; v_y = v_z = 0,$$

where $i = \sqrt{-1}$, ϵ is a small amplitude and ω the frequency of oscillation, the relevant stress component is σ_{xy}, which is often expressed in terms of a complex viscosity η^*, given by

$$\eta^* = \eta' - i\frac{G'}{\omega}.$$

η' is called the dynamic viscosity and G' the dynamic rigidity; both are in general functions of the frequency ω. An alternative and equally popular procedure is to define a 'loss modulus' G'' by means of $G'' = \eta'\omega$ and in this case to refer to G' as the 'storage modulus'.

Finally, in a uniaxial extensional flow given by

$$v_x = \dot{\epsilon}x, \; v_y = -\frac{\dot{\epsilon}}{2}y, \; v_z = -\frac{\dot{\epsilon}}{2}z,$$

where $\dot{\epsilon}$ is a constant extensional strain rate, the corresponding stress distribution is conveniently written in the form

$$\begin{aligned} \sigma_{xx} - \sigma_{yy} &= \sigma_{xx} - \sigma_{zz} = \dot{\epsilon}\eta_E(\dot{\epsilon}), \\ \sigma_{xy} &= \sigma_{xz} = \sigma_{yz} = 0, \end{aligned}$$

where η_E is known as the (uniaxial) extensional viscosity.

Appendix 2

Society of Rheology Bingham Medal Recipients

1948	M Mooney	United States Rubber Company
1949	H Eyring	University of Utah
1950	W F Fair Jnr	Koppers Company
1951	P W Bridgman	Harvard University
1952	A Nadai	Westinghouse Electric Corporation
1953	J D Ferry	University of Wisconsin, Madison
1954	T Alfrey	The Dow Chemical Company
1955	H Leaderman	National Bureau of Standards
1956	A V Tobolsky	Princeton University
1957	C M Zener	Westinghouse Research Laboratory
1958	R S Rivlin	Brown University
1959	E Orowan	Massachusetts Institute of Technology
1960	B Zimm	University of California at San Diego
1961	W R Willets	Titanium Pigment Corportation
1962	W Philippoff	New Jersey Institute of Technology
1963	C A Truesdell	The Johns Hopkins University
1964	J M Burgers	University of Maryland
1965	E Guth	Oak Ridge National Laboratory
1966	P E Rouse	Los Alamos Scientific Laboratory
1967	H Markovitz	Mellon Institute
1968	J L Ericksen	The Johns Hopkins University
1969	S G Mason	McGill University
1970	A Peterlin	Research Triangle Institute
1971	A S Lodge	University of Wisconsin, Madison
1972	R Stein	University of Massachusetts, Amherst
1973	R Simha	Case Western Reserve University
1974	R B Bird	University of Wisconsin, Madison
1975	A N Gent	University of Akron
1976	L E Nielsen	Monsanto Company
1977	A B Metzner	University of Delaware
1978	T L Smith	IBM, Almaden Research Center
1979	W W Graessley	Northwestern University
1980	H Brenner	Massachusetts Institute of Technology
1981	J L White	University of Tennessee
1982	E B Bagley	USDA, Northern Regional Research Center
1983	F R Eirich	Polytechnic Institute of New York
1984	B D Coleman	Rutgers University
1985	R S Porter	University of Massachusetts, Amherst
1986	M M Denn	University of California at Berkeley
1987	C F Curtiss	University of Wisconsin-Madison
1988	W R Schowalter	Princeton University
1989	I M Krieger	Case Western Reserve University
1990	G C Berry	Carnegie Mellon University
1991	L J Zapas	National Bureau of Standards
1992	K F Wissbrun	Hoechst-Celanese Company
1993	D D Joseph	University of Minnesota
1994	A Acrivos	City College of CUNY
1995	D J Plazek	University of Pittsburgh
1996	H H Winter	University of Massachusetts
1997	G G Fuller	Stanford University

Appendix 3

British Society of Rheology Awards

Gold Medal

1966	M Reiner
1969	D Dowson and G R Higginson
1970	G W Scott Blair and V G W Harrison
1972	L R G Treloar
1980	J G Oldroyd
1983	A S Lodge
1984	K Walters FRS
1986	J R A Pearson
1990	Sir Sam Edwards FRS and H Giesekus
1994	M J Crochet

Annual Award

1978	R Whorlow
1979	J F Hutton
1980	E J Hinch FRS
1981	J Meissner, H M Laun H Münstedt and M H Wagner
1982	F M Leslie FRS
1983	D V Boger
1984	K M Beazley
1985	R Buscall
1986	B A Toms
1987	D C-H Cheng
1988	D M Binding
1989	J J Benbow
1990	D M Heyes
1991	P Moldenaers
1992	C J S Petrie
1993	A Evans
1994	G H McKinley
1995	W M Jones

References

Abir D (ed), 1969, *"Contributions of mechanics"*, (Oxford).

Acierno D, Titomanlio G and Greco R, 1974, Chem Eng Sci **29**, 1739-1744.

Acierno D, La Mantia F P, Marrucci G, Titomanlio G, 1976, J non Newtonian Fluid Mechanics **1**, 125-146.

Adams N and Lodge A S, 1964, Phil Trans Roy Soc A**256**, 149-184.

Agoston G A, Harte W H, Hottel H C, Klemm W A, Mysels K J, Pomeroy H H and Thompson J M, 1954, Ind Eng Chem **46**, 1017-1019.

Alfrey T, 1945, Quart J Applied Maths **3**, 143-150.

Alfrey T and Gurnee E F, 1956, *"Rheology: theory and applications"*, Ed F R Eirich, Vol 1, Academic Press, N.Y., 387-429.

Alliluyeva S, 1968, *"Twenty letters to a friend"*, World Books, London.

Amontons G, 1699, Mem de l'Acad Royale **A** (G Kuyper, Amsterdam) 257-282.

Andrade E N da C and Chalmers B, 1932, Proc Roy Soc **A138**, 348-374.

Andrews R D, Hofman-Bang J and Tobolsky A V, 1948, J Polym Sci **3**, 669-692.

Armstrong R C and Ishikawa S, 1980, J Rheol **24**, 143-165.

Ashurst W T and Hoover W G, 1975, Phys Rev A **11**, 658-678.

Astarita G, 1965, I and E C Fundamentals **4**, 354-356.

Astarita G, 1976, J non Newtonian Fluid Mechanics **1**, 203-206.

Astarita G, 1979, J non Newtonian Fluid Mechanics **4**, 285-298.

Astarita G and Marrucci G, 1974, *"Principles of non-Newtonian fluid mechanics"*, Mc-Graw Hill.

Atkins A G and Mai Y-W, 1985, *"Elastic and plastic fracture: metals, polymers, ceramics, composites, biological materials"*, Ellis Horwood.

Atkinson B, Brocklebank M P, Card C C H and Smith J M, 1969, AIChE Journal **15**, 548-553

Avagliano A and Phan-Thien N, 1996, J Fluid Mechanics **312**, 279-298.

Avgousti M and Beris A N, 1993, J non Newtonian Fluid Mechanics **50**, 225-251.

Baaijens F P T, 1994, J non Newtonian Fluid Mechanics **52**, 37-57.

Baaijens F P T, Schoonen J, Peters G W M and Meijer H E H, 1996, Proc 12th Int Cong Rheol Quebec, Eds A Ait-Kadi et al, 419-420.

Bagley E B, 1957, J Appl Phys **28**, 624-627.

Ball R and Richmond P, 1980, J Phys Chem Liquids **9**, 99-116.

Ballman R L, 1965, Rheol Acta **4**, 137-140.

Barnes H A, 1989, J Rheol **33**, 329-366.

Barnes H A, 1995, J non-Newtonian Fluid Mechanics **56**, 221-251.

Barnes H A, 1997, J non Newtonian Fluid Mechanics **70**, 1-33.

Barnes H A, Hutton J F and Walters K, 1989, *"An introduction to rheology"*, Elsevier.

Barus C, 1893, Amer J Sci Series 3, **45**, 87-96. This is an expanded version of a Note in the Proc American Acad, January, 1892.

Batchelor G K, 1970, J Fluid Mechanics **41**, 545-570.

Batchelor G K, 1971, J Fluid Mechanics **46**, 813-829.

Batchelor G K, 1974, Ann Rev Fluid Mech **6**, 227-255.
Batchelor G K, 1975, J Fluid Mechanics **72**, 589-591.
Batchelor G K, 1977, J Fluid Mechanics **83**, 97-117.
Bauer W H and Collins E A, 1967, in *"Rheology: theory and applications"*, Ed F R Eirich, Vol 4, Academic Press, New York, 423-459.
Beard D W, Davies M H and Walters K, 1966, J Fluid Mechanics **24**, 321-334.
Beavers G S and Joseph D D, 1975, J Fluid Mechanics **69**, 475-511.
Beavers G S and Joseph D D, 1977, J Fluid Mechanics **81**, 265-272.
Belhoste B, 1991, *"Augustin-Louis Cauchy: a biography"*. (tr F Ragland) Springer-Verlag, New York.
Bell E T, 1953, *"Men of mathematics"*, Penguin, London.
Benbow J J, Brown R H and Howells E R, 1960, Coll Int Rheologie, Paris.
Benbow J J and Lamb P, 1963, S P E Transactions **3**, No 1, 1-11.
Beris A N, Armstrong R C and Brown R A, 1984, J non-Newtonian Fluid Mechanics **16**, 141-172.
Beris A N, Armstrong R C and Brown R A, 1986, J non Newtonian Fluid Mechanics **19**, 323-347.
Beris A N, Armstrong R C and Brown R A, 1987, J non Newtonian Fluid Mechanics **22**, 129-167.
Berman N S, 1978, Ann Rev Fluid Mech **10**, 47-64.
Bernoulli D, 1738, *"Hydrodynamica"*, Dulsecker, Strasbourg. (See also English translation, Dover, New York 1968.)
Bernstein B, Kadivar M K and Malkus D S, 1981, Comp Methods Appl Mech Eng **27**, 279-302.
Bernstein B, Kearsley A E and Zapas L J, 1963, Trans Soc Rheol **7**, 391-410.
Bernstein B and Malkus D S, 1982, in *"Numerical methods in industrial forming processes"*, Eds J F T Pitman et al, Pineridge Press, 611-620.
Berry J P and Batchelor J, 1968, in *"Polymer systems: deformation and flow"*, Eds R E Wetton and R W Whorlow, McMillan, London, 189-197.
Biller P and Petruccione F, 1990, J Chem Phys **92**, 6322-6331.
Bilmes L, 1942, Nature **142**, 432.
Binding D M, 1988, J non Newtonian Fluid Mechanics **27**, 173-189.
Binding D M and Walters K, 1976, J non Newtonian Fluid Mechanics **1**, 277-286.
Bingham E C, 1916, J Wash Acad Sci **6**, 177-9. See also Nat Bur Stds Bull **13**, 309-353.
Bingham E C, 1922, *"Fluidity and plasticity"*, McGraw-Hill, New York.
Binnington R J and Boger D V, 1985, J Rheol **29**, 887-904.
Bird R B, 1960, Phys Fluids **3**, 539-541, see also **5**, 502.
Bird R B, 1988, Proc 10th Int Cong on Rheol Vol 1, Sydney, Ed P H T Uhlherr, 3-15.
Bird R B, Armstrong R C and Hassager O, 1977a, *"Dynamics of polymeric liquids"*, Vol 1, J Wiley and Sons. Second Edition in 1987.
Bird R B, Curtiss C F, Armstrong R A, Hassager O, 1977b, *"Dynamics of polymeric liquids"* Vol 2 Kinetic theory. J Wiley and Sons. Second Edition in 1987.
Bird R B, Stewart W E and Lightfoot E N, 1960, *"Transport phenomena"*, J Wiley and Sons, New York.
Birkhoff G, 1950, *"Hydrodynamics"*, Dover, New York, p31.

Bland D R, 1960, *"The theory of linear viscoelasticity"*, Pergamon, Oxford.
Bobkowicz A J and Gauvin W H, 1965, Can J Chem Eng **43**, 87-91.
Bobkowicz A J and Gauvin W H, 1967, Chem Eng Sci **22**, 229-241.
Boger D V, 1977/78, J non Newtonian Fluid Mechanics **3**, 87-91.
Boger D V and Denn M M, 1980, J non-Newtonian fluid mechanics **6**, 163-185.
Boger D V and Walters K, 1993, *"Rheological phenomena in focus"*, Elsevier.
Boltzmann L, 1874, Sitzber Kgl Akad Wiss Wien, Math-Naturw Classe **70**, 275-306.
Boltzmann L, 1877, Sitz Kgl Akad Wiss Wien **76**, 815-842.
Boltzmann L, 1878, Wied Ann **5**, 430-432.
Boltzmann L, 1896, *"Vorlesungen über gastheorie"*, Barth, Leipzig.
Bossis G and Brady J F, 1989, J Chem Phys **91**, 1866-1874.
Bossut C, 1797, *"Traité théorique et expérimental d'hydrodynamique"*, 2nd ed, Laron, Paris.
Bousfield D W, Keunings R, Marrucci G and Denn M M,1986, J non-Newtonian Fluid Mechanics **21**, 79-97.
Boyle R, 1660, *"New experiments physico-mechanicall, touching the spring of air and its effects"*, London.
Brady J and Bossis H, 1988, Ann Rev Fluid Mech **20**, 111-157.
Brenner H, 1974, J Multiphase Flow **1**, 195-341.
Brindley G and Broadbent J M, 1973, Rheol Acta **12**, 48-52.
Broadbent J M, Kaye A, Lodge A S and Vale D G, 1968, Nature **217**, 55.
Broda E, 1983, *"Ludwig Boltzmann: man, physicist and philosopher"* (German edt (1955) Deuticke, Berlin), Ox Bow Press, Woodbridge, Conn, USA.
Brodnyan J G, Gaskins F H and Philippoff W, 1957, Trans Soc Rheol **1**, 109-118.
Brodnyan J G, Gaskins F H and Philippoff W, 1958, Trans Soc Rheol **2**, 285-302.
Burgers J M, 1935, in *"First report on viscosity and plasticity"* Chapter 2. Academy of Sciences, Amsterdam, 73-109.
Busse F, 1932, J Phys Chem **34**, 2862-2872.
Butcher J G, 1877, Proc Math Soc Lond **8**, 103-135.

Campbell L and Garnet W, 1884, *"The life of James Clerk Maxwell"*, Macmillan and Co.
Carew E O A, Townsend P and Webster M F, 1993, J non Newtonian Fluid Mechanics **50**, 253-288.
Cauchy A-L, 1827, Ex de Math **2**, 42-56, 60-69, 108-111.
Cauchy A-L, 1828, Ex de Math **3**, 160-187.
Cauchy A-L, 1829, Ex de Math **4**, 293-319.
Chambers Dictionary of Science and Technology, 1971, edited by T C Collocott and W and R Chambers, Edinburgh.
Chan Man Fong C F, 1965, Rheol Acta **4**, 37-39.
Chan Man Fong C F and Walters K, 1965, J Mécanique **4**, 441-453.
Chandrasekhar S, 1943, Rev Mod Phys **15**, 1-89.
Chandrasekhar S, 1961, *"Hydrodynamics and hydromagnetic stability"* Clarendon Press, Oxford.
Chang J C and Denn M M, 1980, in *"Rheology"*, Eds G Astarita et al, Vol 3, Plenum Press, 9-13.

Chang P W, Patten T W and Finlayson B A, 1979, Computers and Fluids **7**, 285-293.
Cheng D C-H, 1973, Nature **245**, 93-95. Expanded version published by the Warren Spring Laboratory (Department of Trade and Industry).
Cheng D C-H and Evans F, 1965, Brit J Appl Phys **16**, 1599-1617.
Chilcott M D and Rallison J M, 1988, J non Newtonian Fluid Mechanics **29**, 381-422.
Choplin L, Carreau P J and AitKadi A, 1983, Polymer Eng and Science **23**, 459-464.
Cogswell F N, 1968, Plast Polym **36**, 109-111.
Cogswell F N, 1969, Rheol Acta **8**, 187-194.
Cogswell F N, 1972a, Polym Eng Sci, **12**, 64-73.
Cogswell F N, 1972b, Trans Soc Rheol **16**, 383-403.
Cogswell F N, 1981, *"Polymer melt rheology"*, 1981, George Goodwin Ltd in association with the Plastics and Rubber Institute.
Colbourn E A, (Ed) 1994, *"Computer simulation of polymers"*, Longman.
Coleman B D, 1962a, Arch Rat Mech Anal **9**, 273-300.
Coleman B D, 1962b,Trans Soc Rheol **6**, 293-300.
Coleman B D and Markovitz H, 1964a, J Appl Phys **35**, 1-9.
Coleman B D and Markovitz H, 1964b, Phys Fluids **7**, 833-841.
Coleman B D, Markovitz H and Noll W, 1966, *"Viscometric flows of non-Newtonian fluids"*, Springer Verlag.
Coleman B D and Noll W, 1959a, Arch Rat Mech Anal **3**, 289-303.
Coleman B D and Noll W, 1959b, J Appl Phys **30**, 1508-1512.
Coleman B D and Noll W, 1960, Arch Rat Mech Anal **6**, 355-370.
Coleman B D and Noll W, 1961, Rev Mod Phys **33**, 239-249.
Cooper M A R, 1997, Notes and Rec, Roy Soc Lond, **51**, 161-174.
Coriolis G, 1830, Ann Chem et Phys **44**, 103-111.
Cotterell B and Kamminga J, 1990, *"Mechanics of pre-industrial technology"*, Cambridge University Press.
Coulomb C A, 1773, Mem Math et Phys **7**, 343-382.
Coulomb C A, 1785, Mem des Savants Etrangers **10**, 161-324.
Coulomb C A, 1801, Mem de l'inst Nat Sci Arts **3**, 246-305.
Court H, Davies A R and Walters K, 1981, J non Newtonian Fluid Mechanics **8**, 95-117.
Cowsley C W, 1970, Univ of Cambridge, Dept of Chem Eng, Polymer Processing Research Centre Report No 6.
Criminale W O, Ericksen J L and Filbey G L, 1958, J Rat Mech Anal **1**, 410-417.
Crochet M J, 1989, Rubber Chemistry and Technology **62**, 426-455.
Crochet M J , Davies A R and Walters K, 1984, *"Numerical simulation of non-Newtonian flow"*, Elsevier.
Crochet M J and Keunings R, 1980, J non Newtonian Fluid Mechanics **7**, 199-212.
Crochet M J and Keunings R, 1982, J non Newtonian Fluid Mechanics **10**, 85-94.
Crochet M J and Pilate G, 1976, J non Newtonian Fluid Mechanics **1**, 247-258.
Crochet M J and Walters K, 1983a, Ann Rev Fluid Mech **15**, 241-260.
Crochet M J and Walters K, 1983b, in *"Computational analysis of polymer processing"*, Eds J R A Pearson and S M Richardson, Applied Science Publ, 21-62.
Crochet M J and Walters K, 1993, Endeavour **17**, No 2, 64-77.
Cruse T A and Rizzo F J, 1968, J Math Anal Appl **22**, 244-259.

Currie P K, 1982, J non Newtonian Fluid Mechanics **11**, 53-68.

D'Alembert J L, 1752, *"Essai d'une nouvelle theorie de la resistance des fluides "*, David, Paris.

Darwish M S, Whiteman J R and Bevis M J, 1992, J non Newtonian Fluid Mechanics **45**, 311-338.

Datta S K, 1964, Phys Fluids **7**, 1915-1919.

Davies R M, 1950, in *"Some recent developments in rheology"*, Ed V G W Harrison, United Trade Press, London, 27-37.

De Gennes P G, 1971, J Chem Phys **55**, 572-579.

De Gennes P G, 1974, *"The physics of liquid crystals"*, Oxford, Clarendon Press.

De Gennes P G, 1979, *"Scaling concepts in polymer physics"*, Cornell University Press, Ithaca.

De Waele A, 1923, Oil Color Chem Assn J **6**, 23-88.

Dealy J M, 1984, J Rheol **28**, 181-185.

Debbaut B and Crochet M J 1986, J non Newtonian Fluid Mechanics **20**, 173-186.

Denn M M, 1980, Ann Rev Fluid Mech **12**, 365-387.

Denn M M, 1990, Ann Rev Fluid Mech **22**, 13-34.

Doi M and Edwards S F, 1978, J Chem Soc Faraday Trans II **74**, 1789-1832.

Doi M and Edwards S F, 1986, *"The theory of polymer dynamics"*, Clarendon Press, Oxford.

Dowson D, 1979, *"History of tribology"*, Longmans, London.

Drazin P G and Reid W H, 1981, *"Hydrodynamic stability"*, Cambridge University Press.

Dryden H L, Murnaghan F D and Bateman H, 1950, *"Hydrodynamics"*, Dover, New York.

Du Buat P, 1816, *"Principes d'hydraulique et de pyrodynamique"*, Firmin Didot, Paris.

Dugas R, 1988, *"A history of mechanics"* (translated by J R Maddox), Dover, New York.

Edwards S F, 1967, Proc Phys Soc **92**, 9-16.

El Kissi N and Piau J M, 1996, *"Rheology for polymer melt processing"*, Eds J M Piau and J F Agassant, Elsevier, 357-420.

Einstein A, 1905, Ann Phys **17**, 549-560.

Eisenschitz R, 1933, Koll Z **64**, 184-195.

Eisenschitz R and Philippoff W, 1933, Naturwissenschaften **21**, 527-528.

Eisenschitz R, Rabinowitsch B and Weissenberg K, 1929, Mittil-dtsch Mat-Pruf Anst **9**, 91-94.

Ericksen J L, 1960, in *"Viscoelasticity: phenomenological aspects"* Ed J T Bergen, Academic Press, New York, 77-92.

Ericksen J L, 1961,Trans Soc Rheol **5**, 22-34.

Ermak D L and McCammon J A, 1978, J Chem Phys **69**, 1352-1360.

Espinasse M, 1956, *"Robert Hooke"* Heinemann, London.

Euler L, 1755, Memoires de Academie de Science Berlin **11**, 217-273.

Evans D J, 1988, Proc Xth Int Cong on Rheol, Sydney, Ed P T Uhlhrerr, 25-30.

Evans D J and Morriss G P, 1988, Phys Rev A **38**, 4142-4148.

Evans D J and Morriss G P, 1990, *"Statistical mechanics of nonequilibrium liquids"*, Academic Press, London.

Eyring H, 1932, Phys Rev **39**, 746-798.

Fabula A G, 1965, Proc 4th Int Cong on Rheology Part 3, Ed E H Lee, Interscience Publishers, New York, 455-479.
Fåhreus R and Lindqvist T, 1931, Amer J Physiol **96**, 562-580.
Fan Y and Crochet M J, 1995, J non-Newtonian Fluid Mechanics **57**, 177-202.
Fano G, 1908, Arch Fisiol **5**, 365-370.
Feitknecht W, 1963, Verh der Schweizerischen Naturfor Gesell **143**, 224-227.
Ferry J D, 1970, *"Viscoelastic properties of polymers"*, J Wiley and Sons, New York.
Finlayson B A, 1972, *"The method of weighted residuals and variational principles"*, Academic Press, New York, 49-81.
Fisher R J and Denn M M, 1975, Chem Eng Sci **30**, 1129-1134.
Fisher R J and Denn M M, 1976, A.I.Ch.E. Journal **22**, 236-246.
Fixman M, 1978, J Chem Phys **69**, 1539-1545.
Flory P J, 1953, *"Principles of polymer chemistry"*, Cornell Univ Press, Ithaca, New York.
Flory P J, 1969, *"Statistical mechanics of chain molecules"*, J Wiley and Sons, New York.
Folgar F and Tucker C L, 1984, J Reinforced Plast Composites **3**, 98-119.
Forrest F and Grierson G A H, 1931, Paper Trade J **92** (22), 39-41.
Fortin A and Zine A, 1992, J non Newtonian Fluid Mechanics **42**, 1-18.
Fortin M and Fortin A, 1989, J non Newtonian Fluid Mechanics **32**, 295-310.
Freundlich H, 1935, *"Thixotropie"*, Hermann and Co, Paris.
Frisch H L and Simha R, 1956, *"Rheology: Theory and application"*, Ed F R Eirich, Vol 1, Academic Press, New York, 525-613.
Fröhlich H and Sack R, 1946, Proc Roy Soc **A185**, 415-430.
Fromm H, 1933, Ing Arch **4**, 432-466.
Fromm H, 1947, ZAMM **26**, 146-150.
Fromm H, 1948, ZAMM **28**, 43-54.
Fuller G G, 1995, *"Optical rheometry of complex fluids"*, Oxford University Press.
Fuller G G, Cathey C A, Hubbard B and Zebrowski B E, 1987, J Rheol **31**, 235-249.

Gadd G E, 1966, Nature **212**, 874-877.
Garner F H, Nissan A H and Wood G F, 1950, Phil Trans Roy Soc Lond A**243**, 37-66.
Gent A N, 1960, Brit J Applied Physics **11**, 165-167.
Gervang B, Davies A R and Phillips T N, 1992, J non-Newtonian Fluid Mechanics **44**, 281-306.
Giesekus H, 1961a, Rheologica Acta **1**, 404-413.
Giesekus H, 1961b, ZAMM **42**, 32-61
Giesekus H, 1961c, ZAMM **42**, 259-262.
Giesekus H, 1962, Rheol Acta **2**, 50-62.
Giesekus H, 1963, Rheol Acta **3**, 59-71.
Giesekus H, 1966, Rheol Acta **5**, 29-35.
Giesekus H, 1968, Rheo Acta 8, 411-421.
Giesekus H, 1972, in *"Progress in heat and mass transfer"*, Eds W R Schowalter et al, Vol 5, Pergamon Press, 187-193.
Giesekus H, 1982, J non Newtonian Fluid Mechanics, **11**, 69-109.

Giesekus H, 1985, in *"Viscoelasticity and rheology"*, Eds A S Lodge, M Renardy and J A Nohel, Academic Press, 157-180.
Giesekus H, 1989, J non Newtonian Fluid Mechanics **33**, 343-348.
Giesekus H, 1990, Rheol Acta **29**, 500-511.
Giesekus H, 1994, *"Phänomenologische rheologie: Eine einführung"*, Springer Verlag.
Giesekus H, 1997, *"Glaubenswagnis. Leben und Erkennen aus der Sicht des Blaise Pascal"*, R Brockhaus Verlag, Wuppertal.
Ginn R F and Denn M M, 1969, A.I.Ch.E. Journal **15**, 450-454.
Ginn R F and Metzner A B, 1969, Trans Soc Rheol **13**, 429-453.
Gjertsen D, 1988, in *"Let Newton be!"* Edited by J Fauvel, R Flood, M Shortland and R Wilson, Oxford University Press, 1-22.
Goddard J D, 1979, Adv Appl Mech **19**, 143-219.
Goldshtik M A, Zametalin V V and Shtern V N, 1982, J Fluid Mechanics **119**, 423-441.
Goldsmith H L and Mason S G, 1967, *"Rheology, theory and applications"*, Ed F R Eirich, Vol 4, Academic Press, New York, 85-250.
Gottlieb D and Orszag S, 1977, *"Numerical analysis of spectral methods"*, SIAM, Philadelphia.
Goublomme A, Draily B and Crochet M J, 1992, J non Newtonian Fluid Mechanics **44**, 171-195.
Grassia P S, Hinch E J and Nitsche L C, 1995, J Fluid Mechanics **282**, 373-403.
Green A E and Rivlin R S, 1957, Arch Rat Mech Anal **1**, 1-21.
Green A E and Rivlin R S, 1960, Arch Rat Mech Anal **4**, 387-404.
Green A E, Rivlin R S and Spencer A J M, 1959, Arch Rat Mech Anal **3**, 82-90.
Green M S and Tobolsky A V, 1946, J Chem Phys **14**, 80-92.
Greensmith H W and Rivlin R S, 1953, Phil Trans Roy Soc Lond A245, 399-429.
Gross B, 1953, *"Mathematical structure of the theories of viscoelasticity"*, Hermann, Paris.
Guénette R and Fortin A, 1995, J non-Newtonian Fluid Mechanics **60**, 27-52.
Guth E and Mark H, 1934, Msh d Chem **65**, 93-121.

Hagen G, 1839, Ann Phys Chem **46**, 423-442.
Hagenbach E, 1860, Ann Phys Chem **10**, 385-425.
Halton P and Scott Blair G W, 1936, J Phys Chem **40**, 561-580.
Harris J (ed), 1973, *"Karl Weissenberg 80th birthday celebration essays"*, East African Lit Bureau, Nairobi.
Harrison V G W, 1954, (ed) Proc 2nd Int Cong on Rheol, Butterworth, London.
Hartley G S, 1938, Nature **142**, 161.
Hatschek E, 1913 Koll Z **13**, 88-96.
Hatzakiriakos S G and Dealy J M, 1991, J Rheol **35**, 497-523.
Hayes J W and Tanner R I, 1965, Proc 4th Int Cong on Rheol, Ed E H Lee, Vol 3, Interscience 389-399.
Helfand E, Wasserman Z R and Weber T A, 1980, Macromolecules **13**, 526-533.
Hencky H, 1923, ZAMM **3**, 241-251.
Hencky H, 1924, ZAMM **4**, 323-334.
Hencky H, 1925, ZAMM **5**, 144-146.
Hencky H, 1929a, Ann Physik (5) **2**, 617-630.

Hencky H, 1929b, Z f Phys **55**, 145-155.
Hermans J J, 1943, Physica **10**, 777-789.
Herschel W and Bulkley R, 1926, Koll Z **39**, 291-300.
Hershey H C and Zakin J L, 1967, I and E C Fund **6**, 381-387.
Herzog R O and Weissenberg K, 1928, Koll Z **46**, 277-289.
Heyes D M, 1986, J Chem Soc Faraday Trans **82**, 1365-1383.
Heyes D M, 1992, Bulletin Brit Soc Rheol **35**, 41-45.
Heyes D M and Melrose J R, 1993, J non Newtonian Fluid Mechanics **46**, 1-28.
Hill R, 1950, *"Plasticity"*, Oxford University Press.
Hill R, 1959, J Mech Phys Sol **7**, 209-225.
Hinch E J and Leal L G, 1975, J Fluid Mechanics **71**, 481-495.
Ho T C and Denn M M, 1977/1978, J non-Newtonian Fluid Mechanics **3**, 179-195.
Hohenemser K and Prager W, 1932, ZAMM **12**, 216-226.
Hooke R, 1665, *"Micrographia"*, J Martyn and J Allestry, London.
Hooke R, 1676, *"A description of helioscopes, and some other instruments"*, London. Printed by T R for John Martyn.
Hooke R, 1678, *"De potentia restitutiva"*, J Martyn, London.
Houwink R, 1937, *"Elasticity, plasticity and the structure of matter"*, Cambridge University Press.
Howells E R and Benbow J J, 1962, Trans and Journal of the Plastics Inst **30**, 240-253.
Hoyt J W, 1975, Coll Int du CNRS No 233,-Polymères et Lubrification, 193-215.
Hoyt J W and Fabula A G, 1964, Navweps Rep 8636 (NOTS TP3670) AD612056.
Hu H H and Joseph D D, 1990, J non Newtonian Fluid Mechanics **37**, 347-377.
Huang X, Phan-Thien N and Tanner R I, 1996, J non Newtonian Fluid Mechanics **64**, 71-92.
Hudson N E and Ferguson J, 1976, Trans Soc Rheol **20**, 265-286.
Huilgol R R and Phan-Thien N, 1997, *"Fluid mechanics of viscoelasticity"*, Elsevier.
Hutton J F, 1965, Proc Roy Soc **A287**, 222-239.
Huygens C, 1690, *"Treatise on light"*, Dover reprint, 1962.

Ignatieff Y A, 1996, *"The mathematical world of Walter Noll. A scientific biography"* Springer.

Jabbarzadeh A, Atkinson J D and Tanner R I, 1997, J non Newtonian Fluid Mechanics, **69**, 169-193.
Jackson K P, Walters K and Williams R W, 1984, J non-Newtonian Fluid Mechanics **14**, 173-188.
Jackson R and Kaye A, 1966, Brit J App Phys **17**, 1355-1360.
James D F, 1966, Nature **212**, 754-756.
James D F and Walters K, 1993, in *"Techniques in rheological measurement"*, Ed A A Collyer, Chapman and Hall, 33-53.
James H M, 1947, J Chem Phys **15**, 651-668.
James H M and Guth E, 1943, J Chem Phys **11**, 455-481.
Janeschitz-Kriegl H, 1983, *"Polymer melt rheology and flow birefringence"*, Springer-Verlag.

Jaumann G, 1905, *"Grundlagen der bewegungslehre"*, Springer, Leipzig.
Jaumann G, 1911, Sitz Akad Wiss Wien (IIa) **120**, 385-530.
Jeffrey G B, 1922, Proc Roy Soc Lond **102**, 161-179.
Jeffreys H, 1929, *"The earth"*, Cambridge University Press.
Jin H, Phan-Thien N and Tanner R I, 1991, Comp Mech **8**, 409-422.
Jobling A and Roberts J E, 1957, *"Rheology of disperse systems"*, edited by C C Mill, Pergamon Press, London, ch 7.
Jobling A and Roberts J E, 1959, J Polymer Sci **36**, 433-441.
Johnson M W Jr, 1960, Phys Fluids **3**, 871-878.
Johnson M W Jr and Segalman D, 1977, J non Newtonian Fluid Mechanics **2**, 255-270.
Jones T E R and Walters K, 1969, Brit J Appl Phys **2**, 815-819.
Joseph D D, 1976a, *"Stability of fluid motions I"* Vol 27, Springer Tracts in Natural Philosophy.
Joseph D D, 1976b, *"Stability of fluid motions II"* Vol 28, Springer Tracts in Natural Philosophy.
Joseph D D, 1981, Arch Rat Mech Anal **75**, 251-256.
Joseph D D, Beavers G S and Fosdick R, 1973, Arch Rat Mech Anal **49**, 381-401.
Joseph D D and Fosdick R, 1973, Arch Rat Mech Anal **49**, 321-380.
Joseph D D, Matta J and Chen K, 1987, J non-Newtonian Fluid Mechanics **24**, 31-65.
Juliusberger F and Pirquet A, 1936, Trans Faraday Soc **32**, 445-452.

Kaeble D H, 1969, J Appl Polymer Science **13**, 2547-2565.
Kalika D S and Denn M M, 1987, J Rheol **31**, 815-834.
Karlsson S K F, Sokolov M and Tanner R I, 1971, Am Inst Chem Eng Symposium Series **67**, 11-20.
Kase S, Matsuo T and Yashimoto Y, 1966, Seni Kikai Gakkaishi **19**, T63-72.
Kawahara M and Takeuchi N, 1977, Comp Fluids **5**, 33-45.
Kaye A, 1962, College of Aeronautics, Cranfield, Note No 134.
Kaye A, 1966, Brit J Appl Phys **17**, 803-806.
Kaye A, Lodge A S and Vale D G, 1968, Rheol Acta **7**, 368-379.
Keentok M V, Georgescu A G, Sherwood A A and Tanner R I, 1980, J non-Newtonian Fluid Mechanics **6**, 303-324.
Kepes A, 1968, Unpublished paper read at 5th Int Congr on Rheol, Kyoto.
Keunings R, 1997, J non Newtonian Fluid Mechanics, **68**, 85-100.
Keynes G 1960, *"A bibliography of Dr Robert Hooke"*, Clarendon Press, Oxford.
Khomami B, Talwar K K and Ganpule H K, 1994, J Rheol **38**, 225-289.
Kim S and Karrila S J, 1991, *"Microhydrodynamics: Principles and selected application"*, Butterworth-Heinemann.
King R C, Apelian M R, Armstrong R C and Brown R A, 1988, J non Newtonian Fluid Mechanics **29**, 147-216.
Kirkwood J G, 1945, J Chem Phys **14**, 51-56.
Kirkwood J G and Riseman J, 1948, J Chem Phys **16**, 565-573.
Kiss G and Porter R S, 1978, J Polym Sci Polym Symp **65**, 193-211.
Klein I and Marshall D I, 1968, *"Computer programs for plastics engineers"*, Rheinhold, New York.

Klein J, 1978, Nature **271**, 143-145.
Klein J and Briscoe B J, 1979, Proc Roy Soc Lond **A365**, 53-73.
Klein O, 1921, Arkiv f mat, astr o fysik **16**, 1-51.
Kohlrausch F W, 1863, Pogg Ann Physik (4) **119**, 337-368.
Kohlrausch F W, 1866, Pogg Ann Physik (5) **128**, 1-19, 207-227, 399-419.
Kohlrausch F W, 1876, Pogg Ann Physik (6) **158**, 337-375.
Kohlrausch R, 1847, Pogg Ann Physik (3) **72**, 353-405.
Kotaka T, Kurata M and Tamura M, 1959, J App Phys **30**, 1705-1712.
Kramers H A, 1944, Physica **11**, 1-19.
Kremer K and Grest G S, 1990, J Chem Phys **92**, 5057-5086.
Krieger I M, 1972, Adv Coll Interface Sci **3**, 111-136.
Krieger I M and Dougherty T J, 1959, Trans Soc Rheol **3**, 137-152.
Kuhn W, 1932, Z Phys Chem **A161**, 16-26, 427-441.
Kuhn W, 1933, Koll Z **62**, 269-285.
Kuhn W, 1934, Koll Z, **68**, 2-15.
Kuhn W, 1936, Koll Z, **76**, 258-271.
Kuhn W and Grün H, 1943, Koll Z **101**, 248-271.
Kuhn W and Grün H, 1946, J Poly Sci **1**, 183-199.
Kuo Y and Tanner R I, 1974, Rheol Acta **13**, 443-456.
Kupffer A T, 1857, Mem Acad St Petersbourg **6**, 397-494.

Lander C H, 1945, J Inst Fuel **19**, 1-11.
Lang H R, 1943, British Rheologists Club Bulletin No 13.
Langevin P, 1908, Comptes Rendus **146**, 530-533.
Larson R G, 1988, *"Constitutive equations for polymer melts and solutions"*, Butterworth, Boston.
Larson R G, 1992, Rheol Acta **31**, 213-263.
Larson R G, Shaqfeh E S G and Muller S J, 1990, J Fluid Mechanics **218**, 573-600.
Laso M and Öttinger H C, 1993, J non Newtonian Fluid Mechanics **47**, 1-20.
Leaderman H, 1943, *"Elastic and creep properties of filamentous materials and other high polymers"*, Textile Foundation, Washington DC.
Lee C-S, Magda J J, De Vries K L and Mays J W, 1992a, Macromolecules **25**, 4744-4750.
Lee C-S, Tripp B C and Magda J J, 1992b, Rheol Acta **31**, 306-308.
Legat V and Marchal J M, 1993, J Rheol **36**, 1325-1348.
Leighton D and Acrivos A, 1987, J Fluid Mechanics **181**, 415-439.
Leonov A I, 1987, J non Newtonian Fluid Mechanics **25**, 1-59.
Leslie F M, 1961, Quart J Mech Applied Math **14**, 36-48.
Leslie F M, 1968, Arch Rat Mech Anal **28**, 265-283.
Leslie F M, 1987, in IMA Vol 5, *"Theory and applications of liquid crystals"* Eds J L Ericksen and D Kinderlehrer, Springer Verlag, New York, 235-254.
Leys S, 1996, *"The analects of Confucius"* (translation by S Leys), W W Norton, New York.
L'Hermite R, 1949, Annls Inst Tech Bâtim No 92, No 5.
Lodge A S, 1951, Proc Camb Phil Soc **47**, 575-584.
Lodge A S, 1954, Proc 2nd Int Cong on Rheol, Ed V G W Harrison, Butterworth, p229.

Lodge A S, 1956, Trans Faraday Soc **52**, 120-130.
Lodge A S, 1964, *"Elastic liquids"*, Academic Press.
Lodge A S, 1973, in *"The Karl Weissenberg 80th birthday celebration essays"*, Ed J Harris, East African Literature Bureau, Kenya, p81-102.
Lodge A S, 1974, *"Body tensor fields in continuum mechanics"*, Academic Press.
Lodge A S and de Vargas L, 1983, Rheol Acta **22**, 151-170.
Lodge A S and Stark J H, 1972, Rheol Acta **11**, 119-126.
Lodge A S and Wu Y, 1971, Rheol Acta **10**, 539-553.
Loring S J, 1950, Trans Amer Soc Mech Engrs **72**, 447-463.
Love A E H, 1952, *"A treatise on the mathematical theory of elasticity"*, 4th Edition, Cambridge University Press.
Lumley J L, 1967, Appl Mech Rev, 1139-1149.
Lumley J L, 1969, Ann Rev Fluid Mech **1**, 367-383.
Lundgren T S, 1971, J Fluid Mechanics **51**, 273-299.
Lunsmann W J, Genieser L, Armstrong R C and Brown R A, 1993, J non Newtonian Fluid Mechanics **48**, 63-100.
Luo X-L, 1996, J non Newtonian Fluid Mechanics **64**, 1973-189.
Luo X-L and Mitsoulis E, 1989, J Rheol **33**, 1307-1327.
Luo X-L and Tanner R I, 1986, J non Newtonian Fluid Mechanics **22**, 61-90.
Luo X-L and Tanner R I, 1988, Int J Num Eng **25**, 9-22.

McKinley G H, Byars J A, Brown R A and Armstrong R C, 1991, J non-Newtonian Fluid Mechanics **40**, 201-229.
Macosko C W and Starita J M, 1971, Soc Plast Eng J **27**, 38-42.
MacSporran W C, 1969, British Society of Rheology Bulletin, No 1, Vol 1, 5-8.
Magda J J and Baek S G, 1994, Polymer **35**, 1187-1194.
Magda J J and Larson R G, 1988, J non-Newtonian Fluid Mechanics **30**, 1-19.
Magda J J, Lou J, Baek S G and De Vries K L, 1991, Polymer **32**, 2000-2009.
Malkus D S and Bernstein B, 1984, J non Newtonian Fluid Mechanics **16**, 77-116.
Marchal J-M and Crochet M J, 1987, J non Newtonian Fluid Mechanics **26**, 77-114.
Marin G, 1988, in *"Rheological measurements"*, Eds A A Collyer and D W Clegg, Elsevier, 297-343.
Markovitz H, 1957, Trans Soc Rheol **1**, 37-52.
Markovitz H, 1968, Physics Today **21**, 23-30.
Markovitz H, 1977, Trans Soc Rheol **21**, 381-398.
Markovitz H, 1985, J Rheol **29**, 777-798.
Markovitz H and Brown D R, 1963, Trans Soc Rheol **7**, 137-154.
Markovitz H and Williamson R B, 1957, Trans Soc Rheol **1**, 25-36.
Markovitz H, Yavorsky P M, Harper R C, Zapas L J and DeWitt T W, 1952, Rev Sci Inst **23**, 430-437.
Marrucci G, 1986, J non Newtonian Fluid Mechanics, **21**, 329-336.
Marrucci G and Grizzuti N, 1986, J non Newtonian Fluid Mechanics, **21**, 319-328.
Marsh B D and Pearson J R A, 1968, Rheol Acta **7**, 326-331.
Martin H C, 1968, Proc 2nd Conf on Matrix Methods in Structural Mechanics, Wright-Patterson A F B, Ohio.

Marvin R S, 1976, Proc 7th Int Cong Rheol, Gothenburg, Sweden, Eds C Klason and J Kubat, 31-36.
Matovich M A and Pearson J R A, 1969, Ind Eng Chem Fund **8**, 512-520.
Maxwell B and Chartoff R P, 1965, Trans Soc Rheol **9**, 41-52.
Maxwell J C, 1853, Trans Roy Soc Edinburgh **20**, 87-120.
Maxwell J C, 1867, Phil Trans Roy Soc Lond **157**, 49-88.
Maxwell J C, 1868, Phil Mag **35**, 129-145, 185-217.
Maxwell J C, 1873, Proc Roy Soc **22**, 46-47.
Maxwell J C, 1878, *"Constitution of solids"* in Encyclopaedia Britannica, 9th edn.
Meissner J, 1969, Rheol Acta **8**, 78-88.
Meissner J, 1972, Trans Soc Rheol **16**, 405-420.
Meissner J, 1975, Pure and Appl Chem **42**, 553-612.
Merrill E W, Smith K A and Chung R Y C, 1966, A I Ch E Journal **12**, 809-810.
Merrington A C, 1943, Nature **152**, 663.
Metzner A B and Park M G, 1964, J Fluid Mechanics **20**, 291-303.
Mewis J, Frith W J, Strivens T A and Russel W B, 1989, AIChE Journal **35**, 415-422.
Meyer K H, 1939, Helv Chim Acta **22**, 1362-1364.
Meyer K H and Ferri C, 1935, Helv Chim Acta **18**, 570-589.
Meyer K H, von Susich G and Valko E, 1932, Koll Z **59**, 208-216.
Meyer O E, 1861, Pogg Ann **113**, 55-86; 193-238.
Meyer O E, 1874, Pogg Ann Physik (6) **151**, 108-119.
Meyer O E, 1878, Wied Ann Physik **4**, 249-267.
Middleman S, 1965, Trans Soc Rheol **9**, 83-93.
Miller C and Goddard J D, 1967, University of Michigan Technical Report (NASA Grant, NSG-659).
Miller M J and Christiansen E B, 1972, A I Ch E Journal **18**, 600-608.
Mooney M, 1931, J Rheol **2**, 210-222.
Mooney M, 1940, J App Phys **11**, 582-592.
Mooney M, 1964, J Appl Phys **35**, 23-26.
Morawetz H, 1995, *"Polymers: the origin and growth of a science"*, Dover, New York.
Morland L W and Lee E H, 1960, Trans Soc Rheol **4**, 233-263.
Münstedt H, 1975, Rheol Acta **14**, 1077-1088.
Murnaghan F D, 1937, Am J Math **59**, 235-260.
Mysels K J, 1971, Chem Eng Progress Symp Series, 111, 45-49.

Naghdi P M, Spencer A J M and England A H, Eds 1994, *"Nonlinear elasticity and theoretical mechanics. In honour of A E Green"*, Oxford University Press.
Nason H K, 1945, J Appl Phys **16**, 338-343.
Natanson L, 1901, Phil Mag (6th Series) **2**, 342-356.
Natanson L, 1903, Bull Acad Sci Cracovie, 767-787.
Navier C L M H, 1821, Ann de Chemie **19**, 244-260.
Navier C L M H, 1822, Bull Soc Philomath, 75-79.
Navier C L M H, 1823, Bull Soc Philomath, 177-181.
Navier C L M H, 1827a Mém Acad Sci Inst France **7**, 375-393.
Navier C L M H, 1827b, Mém Acad Sci Inst France **6**, 389-440.

Needham J, 1959, *"Science and civilisation in China"* Vol 3, Cambridge University Press.
Neesen F, 1874, Ann der Phys u Chemie **153**, 498-516.
Newton I, 1687, *"Philosophiae Naturalis Principia Mathematica"*, (London, 3rd Edn, 1726, translated by Andrew Motte, Cambridge University Press 1934.).
Nickell R E, Tanner R I and Caswell B, 1974, J Fluid Mechanics **65**, 189-206.
Nissan A H, 1994, Letter to D R Oliver 18.3.94 published in Bulletin of British Society of Rheology Vol 37, No 1, 3.
Nitschmann H and Schrade J, 1948, Helv Chim Acta **31**, 297-319.
Noll W, 1955, J Rat Mech Anal **4**, 3-81.
Noll W, 1958, Arch Rat Mech Anal **2**, 197-226.
Noll W, 1962, Arch Rat Mech Anal **11**, 97-105.
Noll W and Virga E, 1990, Arch Rat Mech Anal **111**, 1-31.

Ogden R W, 1984, *"Nonlinear elastic deformations"*, Ellis Horwood, Chichester.
Olabisi O and Williams M C, 1972, Trans Soc Rheol **16**, 727-759.
Oldroyd J G, 1949, Proc 1st Int Cong Rheol, Scheveningen, North Holland, Groningen, II, 130-134.
Oldroyd J G, 1950, Proc Roy Soc A**200**, 523-541.
Oldroyd J G, 1953, Proc Roy Soc A**218**, 122-132.
Oldroyd J G, 1958, Proc Roy Soc A**245**, 278-297.
Oldroyd J G, 1964, *"Second-order effects in elasticity, plasticity and fluid dynamics"*, Ed D Abir, Pergamon, 699-711.
Oldroyd J G, 1965, Proc Roy Soc A**283**, 115-133.
Oldroyd J G, 1976, J Fluid Mechanics **73**, 794-795.
Oldroyd J G, Strawbridge D J and Toms BA, 1951, Proc Phys Soc B**64**, 44-57.
Onoda G Y and Liniger E R, 1990, Phys Rev Letters **64**, 2727-2730.
Onogi S, 1983, British Society of Rheology Bulletin Vol 26.
Oseen C W, 1925, Ark Mat Astr Fys A19, 1-19.
Ostwald W, 1925, Koll Z **36**, 99-117.
Öttinger H C, 1996, *"Stochastic processes in polymeric fluids"*, Springer-Verlag.

Parodi O, 1970, J Phys (Paris) **31**, 581-584.
Parthasarathy M and Klingenberg D J, 1996, Materials Science and Engineering R17, No 2, 57-103.
Pascal B, 1663, *"Traités de l'equilibre des liqueurs et de la pesanteur de la masse de l'air"*, Paris.
Patterson G K, Zakin J L and Rodriguez J M, 1969, Ind and Eng Chem **61**, 22-30.
Pawlowski J, 1954, Koll Z **138**, 6-11.
Pear M R and Weiner J H, 1979, J Chem Phys **71**, 212-224.
Pearson J R A and Petrie C J S, 1965, Proc Fourth Int Congr Rheol, Ed E H Lee, Part 3, J Wiley and Sons, New York, 263-282.
Pearson J R A and Petrie C J S, 1968, in *"Polymer systems: deformation and flow"*, Eds R E Wetton and R W Whorlow, McMillan, London, 163-188.
Perera M G N and Walters K, 1977a, J Non-Newtonian Fluid Mech **2**, 49-81.
Perera M G N and Walters K, 1977b, J Non-Newtonian Fluid Mech **2**, 191-204.

Peterfi T, 1927, Arch Entwicklungsmech, Organ, **112**, 660-695.
Petrie C J S, 1979, *"Elongational flows"*, Pitman.
Petrie C J S and Dealy J M, 1980 in *"Rheology Vol 1 Principles"* Edited by G Astarita, G Marrucci and L Nicolais (Plenum Press) 171-194.
Petrie C J S and Denn M M, 1976, A.I.Ch.E. Journal **22**, 209-236.
Petruccione F and Biller P, 1988a, J Chem Phys **89**, 577-582.
Petruccione F and Biller P, 1988b, Rheol Acta **27**, 557-560.
Phan-Thien N, 1978, J Rheol **22**, 259-283.
Phan-Thien N, 1983, J non-Newtonian Fluid Mechanics **13**, 325-340.
Phan-Thien N, 1985, J non-Newtonian Fluid Mechanics **17**, 37-44.
Phan-Thien N, 1987, J non Newtonian Fluid Mechanics **26**, 327-340.
Phan-Thien N and Tanner R I, 1977, J Non-Newtonian Fluid Mech **2**, 353-365.
Philippoff W, 1934a, Phys Z **35**, 884-989
Philippoff W, 1934b, Phys Z **35**, 900-905
Philippoff W, 1942, *"Viskosität der kolloide"*, T Steinkopff, Leipzig.
Philippoff W, 1956, J Appl Phys **27**, 984-989.
Philippoff W, 1957, Trans Soc Rheol **1**, 95-107.
Philippoff W, 1960a, Trans Soc Rheol **4**, 159-168.
Philippoff W, 1960b, Trans Soc Rheol **4**, 169-177.
Philippoff W, 1961, Trans Soc Rheol **5**, 149-162.
Piau J-M, Bremond M, Couette J M and Piau M, 1994, Rheol Acta **33**, 357-368.
Piau J-M, El Kissi N and Tremblay B, 1990, J non-Newtonian Fluid Mechanics **34**, 145-180.
Pipkin A C, 1964, Rev Mod Phys **36**, 1034-1041.
Pipkin A C and Tanner R I, 1972, Mechanics Today **1**, 262-321.
Planck M, 1917, Sitzber Königl, Preuss Akad Wiss, 323-342.
Poiseuille J L M, 1835, Compt Rend **1**, 554-560.
Poiseuille J L M, 1846, Mem Sav Etranger **9**, 433-544.
Poisson S D, 1829, Mem Acad Sci Inst France **8**, 357-570.
Poynting J H and Thomson J J, 1902, *"Properties of matter"*, London: Charles Griffin & Co Ltd.
Poynting J H, 1913, India-Rubber J, 6.
Prager W, 1961, Quart App Math **18**, 403-407.
Pritchard W G, 1971, Phil Trans Roy Soc Lond A270, 507-556.

Rabinow J, 1948, A.I.E.E. Trans **67**, 1308-1315.
Rabinowitsch B, 1929, Z Phys Chem A145, 1-26.
Radin I, Patterson G K and Zakin J L, 1975, AIChE Journal **21**, 358-371.
Rajagopalan D, Armstrong R C and Brown R A, 1990, J non Newtonian Fluid Mechanics **36**, 159-192.
Rallison J and Hinch E J, 1988, J non-Newtonian Fluid Mechanics **29**, 37-55.
Ramamurthy A V, 1986, J Rheol **30**, 337-358.
Rao P B B, 1964, Appl Sci Res **A14**, 199-210.
Rasmussen H K and Hassager O, 1993, J non Newtonian Fluid Mechanics **46**, 289-305.
Rayleigh Lord, 1879, Proc Roy Soc **29**, 71-97.

Reddy K R and Tanner R I, 1978, J Rheol **22**, 661-665.
Reiner M, 1929a, J Rheol **1**, 5-10.
Reiner M, 1929b, J Rheol **1**, 11-20.
Reiner M, 1945a, J Sci Instr **22**, 127-129.
Reiner M, 1945b, Amer J of Mathematics **67**, 350-362.
Reiner M, 1949, *"Twelve lectures on theoretical rheology"* (North Holland).
Reiner M, 1964, Physics Today **17**, 62.
Reiner M, 1975, *"Selected papers on rheology"* (Amsterdam).
Reiner M, Scott Blair G W and Hawley H B, 1949, J Soc Chem Ind **68**, 327-328.
Reiner M and Scott Blair G W, 1967, in *"Rheology: theory and applications"*, Ed F R Eirich, Vol 4, Academic Press, 461-488.
Renardy M, 1992, Euro J Mech **B11**, 511-516.
Renardy M, 1997a, J non Newtonian Fluid Mechanics **68**, 125-132.
Renardy M, 1997b, J non Newtonian Fluid Mechanics **69**, 99-104.
Reynolds O, 1883a, Proc Roy Soc Lond **35**, 84-99.
Reynolds O, 1883b, Phil Trans Roy Soc Lond **174**, 935-984.
Reynolds O, 1885, Phil Mag **20**, 469-481.
Richardson S, 1971, J Fluid Mechanics **59**, 707-719.
Rigby Z, 1968, British Society of Rheology Bulletin No 4, p23.
Rivlin R S, 1948a, Proc Roy Soc **A193**, 260-281.
Rivlin R S, 1948b, Phil Trans Roy Soc, Lond A**240**, 459-490.
Rivlin R S, 1949a, Proc Camb Phil Soc, **45**, 88-91.
Rivlin R S, 1949b, Trans Faraday Soc **45**, 739-748.
Rivlin R S, 1955, J Rat Mech Anal **4**, 681-702.
Rivlin R S, 1956, J Rat Mech Anal **5**, 179-188.
Rivlin R S, 1959, Physics Today **12**, 32-36.
Rivlin R S, 1965, Proc 4th Int Cong on Rheol Part 2, Ed E H Lee, Interscience Publishers 1,2.
Rivlin R S, 1971, Proc 3rd Canadian Cong of Appl Mech, Ed P Glockner, 39-57.
Rivlin R S, 1984a, Proc 9th Int Cong on Rheol Part 1, Eds B Mena, A Garcia-Rejon and C Rangel-Nafaile, 1-29.
Rivlin R S, 1984b, J non Newtonian Fluid Mech **15**, 253-254.
Rivlin R S, 1996, In *"Collected papers of R S Rivlin"* Eds G I Barenblatt and D D Joseph, Springer-Verlag, xvii-xlvi.
Rivlin R S and Ericksen J L, 1955, J Rat Mech Anal **4**, 323-425.
Rivlin R S and Saunders D W, 1950, Phil Trans Roy Soc Lond A**243**, 251-288.
Rivlin R S and Sawyers K N, 1971, Ann Rev Fluid Mech **3**, 117-146.
Roberts J E, 1952, Unpublished ADE report 13/52, Armament Design Establishment, Knockhold, Kent.
Roberts J E, 1993 Bulletin of British Society of Rheology Vol 36, No 4, 85-89.
Roscoe R, 1950, Brit J Appl Phys **1**, 171-173
Roscoe R, 1965, Brit J Appl Phys **16**, 1567-1571.
Rouse H and Ince S, 1963, *"History of hydraulics"*, Dover, New York, 21.
Rouse P E, 1953, J Chem Phys **21**, 1272-1280.
Russel W B, Saville D A and Schowalter W R, 1989, *"Colloidal dispersions"* Cambridge

University Press.

Sarkar K and Prosperetti A, 1996, J Fluid Mechanics **316**, 223-240.
Sasmal G P, 1995, J non-Newtonian Fluid Mechanics **56**, 15-47.
Savins J G, 1964, Soc Petr Eng J **4**, 203-214.
Savins J G, 1967, Rheol Acta **6**, 323-330.
Savins J G, 1969, in *"Viscous drag reduction"*, Ed C S Wells, Plenum Press, New York, 183-212.
Schalek E and Szegvary A, 1923, Koll Z, **32**, 318-319. English translation provided in W H Bauer and E A Collins, 1967, *"Rheology: theory and applications"*, Ed F R Eirich, Vol 4, Acad Press, New York and London pp 423-459.
Schecter R S, 1961, AIChE Journal **7**, 445-448.
Schofield R K and Scott Blair G W, 1932, Proc Roy Soc **A138**, 707-719.
Schowalter W R, 1978, *"Mechanics of non-Newtonian fluids"*, Pergamon Press.
Schwedoff T, 1889, J Physique [2] **8**, 341-359
Schwedoff T, 1890, J Physique [2] **9**, 34-46.
Scott Blair G W, 1938 *"An introduction to industrial rheology"*, J & A Churchill Ltd.
Scott Blair G W, 1939, British Plastics (March 1939).
Scott Blair G W, 1943, *"A survey of general and applied rheology"*, Pitman, London. Second edition in 1949.
Scott Blair G W, 1953, Nature **172**, 487.
Scott Blair G W, 1968, British Society of Rheology Bulletin No 3.
Scott Blair G W, 1972, Chemistry in Britain **8**, 292-296.
Scott Blair G W, 1976, Rheologica Acta **15**, 365-366.
Scott Blair G W, 1982, Biorheology **19**, 231-235.
Seyer F A and Metzner A B, 1966, Preprints, 6th Naval Hydrodynamics Symposium, Washington DC, p2-1-2-27.
Seyer F A and Metzner A B, 1967, Can J Chem Eng **45**, 121-126.
Shaqfeh E S G, 1996, Ann Rev Fluid Mech **28**, 129-185.
Smith C and Wise M N, 1989, *"Energy and empire"*, Cambridge University Press.
Smith T L, Ferry J D and Schremp F W, 1949, J App Phys **20**, 144-153.
Sokolov M and Tanner R I, 1972, Phys Fluids **15**, 534-539.
Song J H and Yoo J Y, 1987, J non Newtonian Fluid Mechanics **24**, 221-243.
Spencer R S and Dillon R E, 1949, J Colloid Sci **4**, 241-255.
Starita J M, 1980, 'Proc 8th Int Cong on Rheol, Naples, Italy', Eds G Astarita, G Marrucci, G Nicolais, 229-234.
Staudinger H, 1932, *"Die hochmolekularen organischen verbindungen kautschuk und cellulose"*, Springer, Berlin.
Staudinger H, 1933, Trans Farad Soc **29**, 18-32.
Staverman A J and Schwarzl F, 1956, *"Die physik der hochpolymeren"*, Ed H A Stuart, Vol 4, Springer Verlag, Ch 1.
Stein R S and Tobolsky A V, 1948, Text Res J **18**, 302-314.
Stokes G G, 1845, Trans Camb Phil Soc **8**, 287-319.
Stokes G G, 1851, Trans Camb Phil Soc **9**, 8-106.
Sun J, Phan-Thien N and Tanner R I, 1996, J non Newtonian Fluid Mechanics **65**, 75-91.

Szabo B A and Lee G C, 1969, Int J Num Meth in Eng **1**, 301-310.

Takano Y, 1974, Polymer J **6**, 61-71
Tanner R I, 1962, ZAMP, **13**, 573-580.
Tanner R I, 1963, Rheol Acta **3**, 21-26.
Tanner R I, 1969, AIChE Journal **15**, 177-183.
Tanner R I, 1970, J Polym Sci **A8**, 2067-2078.
Tanner R I, 1973, App Poly Symp **20**, 201-208.
Tanner R I, 1975, Trans Soc Rheol **19**, 37-65.
Tanner R I, 1982, J non Newtonian Fluid Mechanics **10**, 169-174.
Tanner R I, 1983, in *"Computational analysis of polymer processing"* Ch 3, eds J R A Pearson and S M Richardson, Applied Science Publishers, London.
Tanner R I, 1988, *"Engineering rheology"*, Oxford University Press, revised edn.
Tanner R I and Jin H, 1991, J non-Newtonian Fluid Mechanics **41**, 171-196.
Tanner R I and Keentok M, 1983, J Rheol **27**, 47-57.
Tanner R I, Nickell R E and Bilger R W, 1975, Comp Meth in Appl Mech and Eng **6**, 155-174.
Tanner R I and Pipkin A C, 1969, Trans Soc Rheol **13**, 471-484.
Tennekes H and Lumley J L, 1972, *"A first course in turbulence"*, MIT Press, Cambridge Mass.
Thomas R H and Walters K, 1963, Proc Roy Soc A274, 371-385.
Thomas R H and Walters K 1964a, J Fluid Mechanics **18**, 33-43.
Thomas R H and Walters K 1964b, J Fluid Mechnics **19**, 557-560.
Thomas T Y, 1955, Proc Nat Acad Sci USA **41**, 762-770.
Thomson J J, 1888, *"Applications of dynamics to physics and chemistry"*, Macmillan, London, Chapter 8.
Thomson W, 1865, Proc Roy Soc **14**, 289-297.
Thomson W, 1878, *"Elasticity and heat"*, in Encyclopaedia Britannica, 9th edn.
Timoshenko S, 1953, *"History of strength of materials"*, McGraw-Hill, New York.
Tomita Y, 1959, Bull Japan Soc Mech Eng **2**, 469-474.
Toms B A, 1949, Proc 1st Int Congr Rheol, Scheveningen, N Holland, Groningen, pp II 135-141.
Toms B A, 1977, Phys Fluids **20**, 53-55.
Tordella J P, 1956, J Appl Phys **27**, 454-458.
Tordella J P, 1957, Trans Soc Rheol **1**, 203-212.
Tordella J P, 1958, Rheol Acta **1**, 216-221.
Tordella J P, 1963, J Appl Polym Sci **7**, 215-229.
Townsend P, 1973, Rheol Acta **12**, 13-18.
Townsend P, 1984, J non-Newtonian Fluid Mechanics **14**, 265-278.
Tran-Cong T and Phan-Thien N, 1988, J non Newtonian Fluid Mechanics **30**, 37-46.
Treloar L R G, 1943, Trans Farad Soc **39**, 36-41.
Treloar L R G, 1958, *"The physics of rubber elasticity"*, Oxford University Press, 2nd Edn.
Tresca H, 1864, Compt Rendus **59**, 754-758.
Tresca H, 1872, Mem par Div Sav, Acad Sci Paris **20**, 75-135.

Trouton F T, 1906, Proc Roy Soc A**77**, 426-440.
Truesdell C, 1952, J Rat Mech Anal **1**, 125-300.
Truesdell C, 1955, J Rat Mech Anal **4**, 83-133, 1019-1020.
Truesdell C, 1960a, *"Principles of continuum mechanics"*, Socony Mobil Oil Co In, Dallas, Texas, 188.
Truesdell C, 1960b, Trans Soc Rheol **4**, 9-29.
Truesdell C, 1964, *"Second-order effects in elasticity, plasticity and fluid dynamics"*, Ed D Abir, Pergamon, 1-47.
Truesdell C, 1965a, Ed *"Continuum mechanics II The rational mechanics of materials"*, Gordon and Breach.
Truesdell C, 1965b, Proc 4th Int Cong on Rheol Part 2, Ed E H Lee, Interscience Publishers, 3-30.
Truesdell C, 1968, *"Essays in the history of mechanics"*, Springer Verlag.
Truesdell C, 1980, *"Rheology Vol 1 Principles"*, Eds G Astarita, G Marrucci and L Nicolais, Plenum Press, 1-27.
Truesdell C, 1984, J non Newtonian Fluid Mech **15**, 249-251.
Truesdell C and Noll W, 1965, *"The non-linear field theories of mechanics"*, in Handbuch der Physik III/3, S Fluegge (ed), Springer Verlag.
Truesdell C and Toupin R, 1960, *"The classical field theories"* in Handbuch der Physik III/3, S Flügge,(ed) Berlin, Springer Verlag III/1, 226-693.
Tschoegl N W, 1989, *"The phenomenological theory of linear viscoelasticity"*, Springer Verlag.
Turner M J, Clough R W, Martin H C and Topp L P, 1956, J Aero Sci **23**, 805-823.

Valson C A, 1868, *"La vie et les travaux du Baron Cauchy"*, Gauthier-Villars, Paris.
Vest C M and Arpaci V S, 1969, J Fluid Mechanics **36**, 613-623.
Vinogradov G V and Malkin A Ya, 1980, *"Rheology of polymers"*, Mir, Moscow.
Vinogradov G V and Malkin A Ya, Yanovskii Yu G, Borisenkova E K, Yarliykov B V and Berezhnaya G V, 1972, J Polymer Science Part A-2 **10**, 1061-1084.
Viriyayuthakorn M and Caswell B, 1980 J non Newtonian Fluid Mechanics **6**, 245-267.
Virk P S, 1975, AIChE Journal **21**, 625-656.
Virk P S, Merrill E W, Mickley H S and Smith K A, 1967, J Fluid Mechanics **30**, 305-328.
Visscher P B and Heyes D M, 1994, J Chem Phys **101**, 6096-6100.
Voigt W, 1889, Göttinger Abh **36**, No 1.
Voigt W, 1892, Ann Physik (2) **47**, 671-693.
von Smoluchowski M, 1906, Ann Phys **21**, 756-780.
von Smoluchowski M, 1915, Ann Phys **48**, 1103-1112.

Wagner M H, 1978, J non Newtonian Fluid Mechanics, **4**, 39-55.
Wales J L S, 1976, *"The application of flow birefringence to rheological studies of polymer melts"*, Delft University Press.
Wali K C, 1991, *"Chandra: a biography of S Chandrasekhar"*, Viking Penguin, India, New Delhi.
Wall F T, 1943, J Chem Phys **11**, 67-71, 527-530.
Walters K, 1962, J Mécanique **1**, 479-486.

Walters K, 1975, *"Rheometry"* Chapman and Hall.
Walters K, 1979, J non Newtonian Fluid Mechanics **5**, 113-124.
Walters K, 1989, Rheol Acta **28**, 1.
Walters K and Barnes H A, 1980 in *"Rheology Vol 1 Principles"* Edited by G Astarita, G Marrucci and L Nicolais (Plenum Press) 171-194.
Walters K and Tanner R I, 1992, in *"Transport processes in bubbles, drops and particles"* Eds Chhabra R R and DeKee D, Hemisphere Publ Corp, New York, 73-86.
Warburg E, 1870, Pogg Ann **140**, 367-384.
Warner H, 1972, Ind Eng Chem Fund **11**, 379-387.
Wasserman S H and Graessley W W, 1996, Polymer Eng and Sci **36**, 852-861.
Weber W, 1835, Ann Phys Chem **34**, 247-257.
Weber W, 1841, Ann Phys Chem **54**, 1-18.
Weissenberg K, 1931, Abhg Preuss Akad Wiss No **2**, 617-630.
Weissenberg K, 1934, Arch Sci Phys et Nat (5) **17**, 44-106.
Weissenberg K, 1947, Nature **159**, 310-311.
Weissenberg K, 1949, Proc 1st Int Congr on Rheology, **2**, 114-118.
Wertheim G, 1847, Ann de Chemie **21**, 385-414.
Wertheim G, 1851, Ann de Chemie **31**, 19-36.
Westfall R S, 1980, *"Never at rest"*, Cambridge University Press.
Wheeler J A and Wissler E, 1965, AIChE Journal **11**, 207-212, 212-216.
Wheeler J A and Wissler E, 1966, Trans Soc Rheol **10**, 353-367.
Whetham W C D, 1890, Proc Roy Soc **48**, 225-230.
White A, 1966, Nature **211**, 1390.
White A, 1967a, Nature **214**, 585-586.
White A, 1967b, Nature **216**, 994-995.
White J L, 1964, J Appl Poly Sci **8**, 2339-2357.
White J L, 1990, *"Principles of polymer engineering rheology"*, J Wiley and Sons.
White J L and Metzner A B, 1963, J Appl Polym Sci **7**, 1867-1889.
White J L and Tokita N, 1967, J Phys Soc Japan **22**, 719-724.
Wiechert E, 1893, Wied Ann Phys Lpz **50**, 335-348, 546-570.
Wiedemann G, 1856, Ann Phys Chem **99**, 177-233.
Wiegel F W, 1969, Physica **42**, 156-164.
Wiegel F W and de Bats F T 1969, Physica **42**, 156-164.
Wiley F E, 1941, Ind Eng Chem **33**, 1377-1380.
Williams M L, Landel R F and Ferry J D, 1955, J Am Chem Soc **77**, 3701-3707.
Winslow W M, 1947, US Patent 2, 417, 850.
Winslow W M, 1949, J Appl Phys **20**, 1137-1140.
Winter H H, 1989, Rheologica Acta **28**, 437-448.
Wöhlisch E, 1926, Z Biol **85**, 379-390.
Wöhlisch E, 1939, Helv Chim Acta **22**, 1358-1362.
Wookey N, 1961, British Society of Rheology Bulletin No 3.

Xue S-C, Phan-Thien N and Tanner R I, 1995, J non Newtonian Fluid Mechanics **59**,

191-213.

Yamamoto M, 1956, J Phys Soc Japan **11**, 413-421.
Yamamoto M, 1957, J Phys Soc Japan **12**, 1148-1158.
Yamamoto M, 1958, J Phys Soc Japan **13**, 1200-1211.
Yamamoto M and Inagaki H, 1952, Busseirion Kankyu No 55, p26.
Yoo J Y and Joseph D D, 1985, J Non-Newtonian Fluid Mechanics **19**, 15-42.
Yoo J Y and Na Y, 1991, J non Newtonian Fluid Mechanics **39**, 89-106.
Young D M and Wheeler M F, 1964, in *"Nonlinear problems in engineering"*, Ed W F Ames, Academic Press, New York, 220-246.
Young T, 1807, *"A course of lectures on natural philosophy and the mechanical arts"*, Taylor and Walton, London.

Zaremba S K, 1903, Bull Acad Sci Cracovie, 85-93, 380-403, 594-614.
Zaremba S K, 1937, Mém Sci Math No 82, Gauthier-Villars, Paris.
Zimm B H, 1956, J Chem Phys **24**, 269-278.

Author Index

Subject Index

Kolibri 360 GmbH
Ferdinand-Jühlke-Straße 7
99095 Erfurt, Deutschland
produktsicherheit@kolibri360.de